AF713166

Excel for Statistics

Excel for Statistics is a series of textbooks that explain how to use Excel to solve statistics problems in various fields of study. Professors, students, and practitioners will find these books teach how to make Excel work best in their respective field. Applications include any discipline that uses data and can benefit from the power and simplicity of Excel. Books cover all the steps for running statistical analyses in Excel 2016, Excel 2013, Excel 2010 and Excel 2007. The approach also teaches critical statistics skills, making the books particularly applicable for statistics courses taught outside of mathematics or statistics departments.

Series editor: Thomas J. Quirk

The following books are in this series:

T.J. Quirk, *Excel 2016 for Business Statistics: A Guide to Solving Practical Problems*, Excel for Statistics. Springer International Publishing Switzerland 2016.

T.J. Quirk, *Excel 2016 for Engineering Statistics: A Guide to Solving Practical Problems*, Excel for Statistics. Springer International Publishing Switzerland 2016.

T.J. Quirk, M. Quirk, H.F. Horton, *Excel 2016 for Biological and Life Sciences Statistics: A Guide to Solving Practical Problems*, Excel for Statistics. Springer International Publishing Switzerland 2016.

T.J. Quirk, E. Rhiney, *Excel 2016 for Marketing Statistics: A Guide to Solving Practical Problems*, Excel for Statistics. Springer International Publishing Switzerland 2016.

T.J. Quirk. *Excel 2016 for Educational and Psychological Statistics: A Guide to Solving Practical Problems*, Excel for Statistics. Springer International Publishing Switzerland 2016.

T.J. Quirk, *Excel 2016 for Social Science Statistics: A Guide to Solving Practical Problems*, Excel for Statistics. Springer International Publishing Switzerland 2016.

T.J. Quirk, M. Quirk, H. Horton, *Excel 2016 for Physical Sciences Statistics: A Guide to Solving Practical Problems*. Excel for Statistics. Springer International Publishing Switzerland 2016.

T.J. Quirk, S. Cummings, *Excel 2016 for Health Services Management Statistics: A Guide to Solving Practical Problems*. Excel for Statistics. Springer International Publishing Switzerland 2016.

T.J. Quirk, J. Palmer-Schuyler, *Excel 2016 for Human Resource Management Statistics: A Guide to Solving Practical Problems*, Excel for Statistics. Springer International Publishing Switzerland 2016.

T.J. Quirk, M. Quirk, H.F. Horton. *Excel 2016 for Environmental Sciences Statistics: A Guide to Solving Practical Problems*, Excel for Statistics. Springer International Publishing Switzerland 2016.

T.J. Quirk, J. Palmer-Schuyler, *Excel 2013 for Human Resource Management Statistics: A Guide to Solving Practical Problems*, Excel for Statistics. Springer International Publishing Switzerland 2016.

T.J. Quirk, S. Cummings, *Excel 2013 for Health Services Management Statistics: A Guide to Solving Practical Problems*, Excel for Statistics. Springer International Publishing Switzerland 2016.

T.J. Quirk, M. Quirk, H. Horton, *Excel 2013 for Physical Sciences Statistics: A Guide to Solving Practical Problems*, Excel for Statistics. Springer International Publishing Switzerland 2016.

T.J. Quirk, J. Palmer-Schuyler, *Excel 2010 for Human Resource Management Statistics: A Guide to Solving Practical Problems*, Excel for Statistics. Springer International Publishing Switzerland 2014.

T.J. Quirk, *Excel 2013 for Business Statistics: A Guide to Solving Practical Problems*, Excel for Statistics. Springer International Publishing Switzerland 2015.

T.J. Quirk, M. Quirk, H.F. Horton, *Excel 2013 for Biological and Life Sciences Statistics: A Guide to Solving Practical Problems*, Excel for Statistics. Springer International Publishing Switzerland 2015.

T.J. Quirk, *Excel 2013 for Social Science Statistics: A Guide to Solving Practical Problems*, Excel for Statistics. Springer International Publishing Switzerland 2015.

T.J. Quirk, *Excel 2013 for Engineering Statistics: A Guide to Solving Practical Problems*, Excel for Statistics. Springer International Publishing Switzerland 2015.

T.J. Quirk, *Excel 2013 for Educational and Psychological Statistics: A Guide to Solving Practical Problems*, Excel for Statistics. Springer International Publishing Switzerland 2015.

T.J. Quirk, M. Quirk, H.F. Horton, *Excel 2013 for Environmental Sciences Statistics: A Guide to Solving Practical Problems*, Excel for Statistics. Springer International Publishing Switzerland 2015.

T.J. Quirk, M. Quirk, H.F. Horton, *Excel 2010 for Environmental Sciences Statistics: A Guide to Solving Practical Problems*, Excel for Statistics. Springer International Publishing Switzerland 2015.

Additional Statistics books by Dr. Tom Quirk that have been published by Springer

T.J. Quirk, *Excel 2010 for Engineering Statistics: A Guide to Solving Practical Problems*. Springer International Publishing Switzerland 2014.

T.J. Quirk, S. Cummings, *Excel 2010 for Health Services Management Statistics: A Guide to Solving Practical Problems*. Springer International Publishing Switzerland 2014.

T.J. Quirk, M. Quirk, H. Horton, *Excel 2010 for Physical Sciences Statistics: A Guide to Solving Practical Problems*. Springer International Publishing Switzerland 2013.

T.J. Quirk, M. Quirk, H.F. Horton, *Excel 2010 for Biological and Life Sciences Statistics: A Guide to Solving Practical Problems*. Springer Science+Business Media New York 2013.

T.J. Quirk, M. Quirk, H.F. Horton, *Excel 2007 for Biological and Life Sciences Statistics: A Guide to Solving Practical Problems*. Springer Science+Business Media New York 2013.

T.J. Quirk, *Excel 2010 for Social Science Statistics: A Guide to Solving Practical Problems*. Springer Science+Business Media New York 2012.

T.J. Quirk, *Excel 2010 for Educational and Psychological Statistics: A Guide to Solving Practical Problems*. Springer Science+Business Media New York 2012.

T.J. Quirk, *Excel 2007 for Business Statistics: A Guide to Solving Practical Problems*. Springer Science +Business Media New York 2012.

T.J. Quirk, *Excel 2007 for Social Science Statistics: A Guide to Solving Practical Problems*. Springer Science+Business Media New York 2012.

T.J. Quirk, *Excel 2007 for Educational and Psychological Statistics: A Guide to Solving Practical Problems*. Springer Science+Business Media New York 2012.

T.J. Quirk, *Excel 2010 for Business Statistics: A Guide to Solving Practical Problems*. Springer Science +Business Media 2011.

More information about this series at http://www.springer.com/series/13491

Thomas J. Quirk

Excel 2016 for Social Science Statistics

A Guide to Solving Practical Problems

Springer

Thomas J. Quirk
Webster University
St. Louis, MO, USA

Excel for Statistics
ISBN 978-3-319-39710-8 ISBN 978-3-319-39711-5 (eBook)
DOI 10.1007/978-3-319-39711-5

Library of Congress Control Number: 2016941691

© Springer International Publishing Switzerland 2016
This work is subject to copyright. All rights are reserved by the Publisher, whether the whole or part of the material is concerned, specifically the rights of translation, reprinting, reuse of illustrations, recitation, broadcasting, reproduction on microfilms or in any other physical way, and transmission or information storage and retrieval, electronic adaptation, computer software, or by similar or dissimilar methodology now known or hereafter developed.
The use of general descriptive names, registered names, trademarks, service marks, etc. in this publication does not imply, even in the absence of a specific statement, that such names are exempt from the relevant protective laws and regulations and therefore free for general use.
The publisher, the authors and the editors are safe to assume that the advice and information in this book are believed to be true and accurate at the date of publication. Neither the publisher nor the authors or the editors give a warranty, express or implied, with respect to the material contained herein or for any errors or omissions that may have been made.

This Springer imprint is published by Springer Nature
The registered company is Springer International Publishing AG Switzerland

This book is dedicated to the more than 3,000 students I have taught at Webster University's campuses in St. Louis, London, and Vienna; the students at Principia College in Elsah, Illinois; and the students at the Cooperative State University of Baden-Wuerttemburg in Heidenheim, Germany. These students taught me a great deal about the art of teaching. I salute them all, and I thank them for helping me to become a better teacher.

Thomas J. Quirk

Preface

Excel 2016 for Social Science Statistics: A Guide to Solving Practical Problems is intended for anyone looking to learn the basics of applying Excel's powerful statistical tools to their social science courses or work activities. If understanding statistics isn't your strongest suit, you are not especially mathematically inclined, or if you are wary of computers, then this is the right book for you.

Here you'll learn how to use key statistical tests using Excel without being overpowered by the underlying statistical theory. This book clearly and methodically shows and explains how to create and use these statistical tests to solve practical problems in the social sciences.

Excel is an easily available computer program for students, instructors, and managers. It is also an effective teaching and learning tool for quantitative analyses in social science courses. The powerful numerical computational ability and the graphical functions available in Excel make learning statistics much easier than in years past. However, this is the first book to show Excel's capabilities to more effectively teach social science statistics; it also focuses exclusively on this topic in an effort to render the subject matter not only applicable and practical, but also easy to comprehend and apply.

Unique features of this book:

- You will be told each step of the way, not only *how* to use Excel, but also *why* you are doing each step so that you can understand what you are doing and not merely learn how to use statistical tests by rote.
- Includes specific objectives embedded in the text for each concept, so you can know the purpose of the Excel steps.
- Includes 167 color screenshots so that you can be sure you are performing the Excel steps correctly.
- This book is a tool that can be used either by itself or along with *any* good statistics book.
- Practical examples and problems are taken from the social sciences, including political science, sociology, anthropology, education, and psychology.

- Statistical theory and formulas are explained in clear language without bogging you down in mathematical fine points.
- You will learn both how to write statistical formulas using Excel and how to use Excel's drop-down menus that will create the formulas for you.
- This book does not come with a CD of Excel files which you can upload to your computer. Instead, you'll be shown how to create each Excel file yourself. In a work situation, your colleagues will not give you an Excel file; you will be expected to create your own. This book will give you ample practice in developing this important skill.
- Each chapter presents the steps needed to solve a practical social science problem using Excel. In addition, there are three practice problems at the end of each chapter so you can test your new knowledge of statistics. The answers to these problems appear in Appendix A.
- A "Practice Test" is given in Appendix B to test your knowledge at the end of the book. The answers to these practical social science problems appear in Appendix C.

This book is appropriate for use in any course in Social Science Statistics (at both undergraduate and graduate levels) as well as for managers who want to improve the usefulness of their Excel skills.

This book has a single author, Dr. Tom Quirk, a current Professor of Marketing at the George Herbert Walker School of Business and Technology at Webster University in St. Louis, Missouri (USA), where he teaches Marketing Statistics, Marketing Research, and Pricing Strategies. The ideas in this book have been thoroughly tested in Professor Quirk's Marketing Statistics and Marketing Research courses. At the beginning of his academic career, Prof. Quirk spent 6 years in educational research at the American Institutes for Research and Educational Testing Service. He then taught Social Psychology, Educational Psychology, and General Psychology at Principia College in Elsah, Illinois (USA). He has published articles in *the Journal of Educational Psychology*, *Journal of Educational Research*, *Review of Educational Research*, *Journal of Educational Measurement*, *Educational Technology*, *The Elementary School Journal*, *Journal of Secondary Education*, *Educational Horizons*, *and Phi Delta Kappan*. In addition, Professor Quirk has written more than 60 textbook supplements in Management and Marketing, published more than 20 articles in professional journals, and presented more than 20 papers at professional meetings, including annual meetings of the American Educational Research Association, the American Psychological Association, and the National Council on Measurement in Education. He holds a B.S. in Mathematics from John Carroll University, both an M.A. in Education and a Ph.D. in Educational Psychology from Stanford University, and an M.B.A. from the University of Missouri-St. Louis.

St. Louis, MO Thomas J. Quirk

Acknowledgments

Excel 2016 for Social Science Statistics: A Guide to Solving Practical Problems is the result of inspiration from three important people: my two daughters and my wife. Jennifer Quirk McLaughlin invited me to visit her M.B.A. classes several times at the University of the Witwatersrand in Johannesburg, South Africa. These visits to a first-rate M.B.A. program convinced me there was a need for a book to teach students how to solve practical social science problems using Excel. Meghan Quirk-Horton's dogged dedication to learning the many statistical techniques needed to complete her Ph.D. dissertation illustrated the need for a statistics book that would make this daunting task more user-friendly. And Lynne Buckley-Quirk was the number one cheerleader for this project from the beginning, always encouraging me and helping me remain dedicated to completing it.

Marc Strauss, my editor at Springer, caught the spirit of this idea in our first phone conversation and shepherded this book through the idea stages until it reached its final form. His encouragement and support were vital to this book seeing the light of day. I thank him for being such an outstanding product champion throughout this process. And Christine Crigler at Springer did her usual first-rate job in coordinating the editing and production of this book; she is always a pleasure to work with.

Contents

Chapter 1
Sample Size, Mean, Standard Deviation, and Standard Error of the Mean

This chapter deals with how you can use Excel to find the average (i.e., "mean") of a set of scores, the standard deviation of these scores (STDEV), and the standard error of the mean (s.e.) of these scores. All three of these statistics are used frequently and form the basis for additional statistical tests.

1.1 Mean

The *mean* is the "arithmetic average" of a set of scores. When my daughter was in the fifth grade, she came home from school with a sad face and said that she didn't get "averages." The book she was using described how to find the mean of a set of scores, and so I said to her:

> "Jennifer, you add up all the scores and divide by the number of numbers that you have."
> She gave me "that look," and said: "Dad, this is serious!" She thought I was teasing her. So I said:
> "See these numbers in your book; add them up. What is the answer?" (She did that.)
> "Now, how many numbers do you have?" (She answered that question.)
> "Then, take the number you got when you added up the numbers, and divide that number by the number of numbers that you have."

She did that, and found the correct answer. You will use that same reasoning now, but it will be much easier for you because Excel will do all of the steps for you.

We will call this average of the scores the "mean" which we will symbolize as: $\overline{X}$, and we will pronounce it as: "Xbar."

The formula for finding the mean with your calculator looks like this:

$$\overline{X} = \frac{\Sigma X}{n} \tag{1.1}$$

© Springer International Publishing Switzerland 2016

T.J. Quirk, *Excel 2016 for Social Science Statistics*, Excel for Statistics,
DOI 10.1007/978-3-319-39711-5_1

The symbol Σ is the Greek letter sigma, which stands for "sum." It tells you to add up all the scores that are indicated by the letter X, and then to divide your answer by n (the number of numbers that you have).

Let's give a simple example:

Suppose that a political scientist developed a survey measuring the attitudes of registered voters on a variety of issues and that one of the items on this survey was the following: "Wealthy people should pay taxes at a much higher rate than poor people." Suppose, further, that a 7-point rating scale was used for this item such that 1 = strongly disagree, and 7 = strongly agree.

Suppose that you had these six ratings on this one item:

6
4
5
3
2
5

To find the mean of these scores, you add them up, and then divide by the number of scores. So, the mean is: 25/6 = 4.17 (close to the middle of the 7-point scale).

1.2 Standard Deviation

The *standard deviation* tells you "how close the scores are to the mean." If the standard deviation is a small number, this tells you that the scores are "bunched together" close to the mean. If the standard deviation is a large number, this tells you that the scores are "spread out" a greater distance from the mean. The formula for the standard deviation (which we will call STDEV) and use the letter, S, to symbolize is:

$$\text{STDEV} = S = \sqrt{\frac{\Sigma(X - \overline{X})^2}{n - 1}} \tag{1.2}$$

The formula look complicated, but what it asks you to do is this:

1. Subtract the mean from each score $(X - \overline{X})$.
2. Then, square the resulting number to make it a positive number.
3. Then, add up these squared numbers to get a total score.
4. Then, take this total score and divide it by n – 1 (where n stands for the number of numbers that you have).
5. The final step is to take the square root of the number you found in step 4.

You will not be asked to compute the standard deviation using your calculator in this book, but you could see examples of how it is computed in any basic statistics book. Instead, we will use Excel to find the standard deviation of a set of scores. When we use Excel on the six numbers we gave in the description of the mean above, you will find that the *STDEV* of these numbers, S, is 1.47.

1.3 Standard Error of the Mean

The formula for the *standard error of the mean* (*s.e.*, which we will use $S_{\overline{X}}$ to symbolize) is:

$$\text{s.e.} = S_{\overline{X}} = \frac{S}{\sqrt{n}} \tag{1.3}$$

To find ***s.e.***, all you need to do is to take the standard deviation, STDEV, and divide it by the square root of n, where n stands for the "number of numbers" that you have in your data set. In the example under the standard deviation description above, the *s.e.* = 0.60. (You can check this on your calculator).

If you want to learn more about the standard deviation and the standard error of the mean, see Weiers (2011) and Neuman (2000).

Now, let's learn how to use Excel to find the sample size, the mean, the standard deviation, and the standard error or the mean using a geometry test given to a class of 9th graders at the end of the first term of the school year (50 points possible). The hypothetical data appear in Fig. 1.1.

Fig. 1.1 Worksheet Data for a Geometry Test (Practical Example)

Student	Geometry Test Score
1	10
2	10
3	12
4	16
5	22
6	29
7	39
8	47

1.4 Sample Size, Mean, Standard Deviation, and Standard Error of the Mean

Objective: To find the sample size (n), mean, standard deviation (STDEV), and standard error of the mean (s.e.) for these data

Start your computer, and click on the Excel 2016 icon to open a blank Excel spreadsheet.

Click on: Blank Workbook

Enter the data in this way:

A3: Student
B3: Geometry Test Score
A4: 1

1.4.1 Using the Fill/Series/Columns Commands

Objective: To add the student numbers 2-8 in a column underneath student #1

Put pointer in A4
Home (top left of screen)

Important note: The "Paste" command should be on the top of your screen on the far left of the screen.

Important note: Notice the Excel commands at the top of your computer screen: File → Home → Insert → Page Layout → Formulas → etc.

If these commands ever "disappear" when you are using Excel, you need to click on "Home" at the top left of your screen to make them re-appear!

Fill (top right of screen: click on the down arrow; see Fig. 1.2)

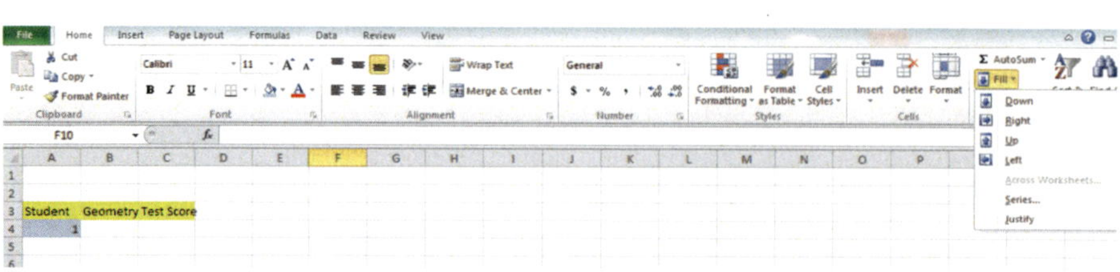

Fig. 1.2 Home/Fill/Series commands

Series
Columns
Step value: 1
Stop value: 8 (see Fig. 1.3)

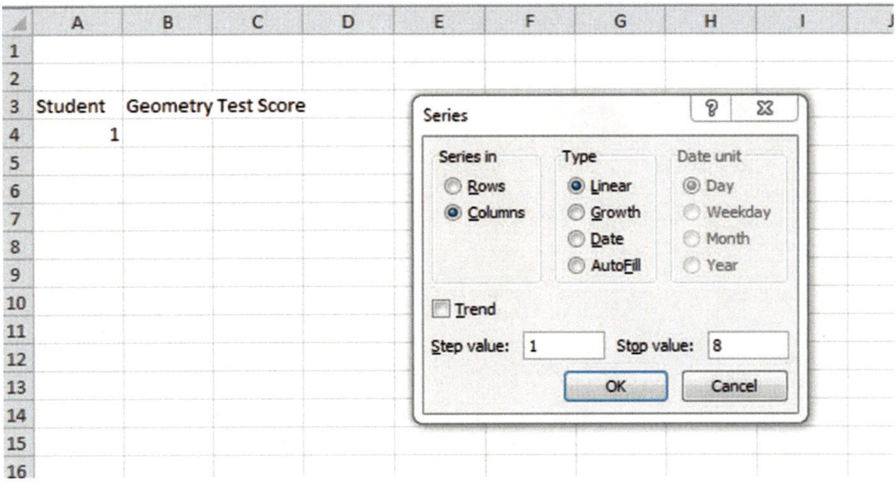

Fig. 1.3 Example of Dialogue Box for Fill/Series/Columns/Step Value/Stop Value commands

OK

The student numbers should be identified as 1–8, with 8 in cell A11.

Now, enter the Geometry Test Scores in cells B4:B11.

Since your computer screen shows the information in a format that does not look professional, you need to learn how to "widen the column width" and how to "center the information" in a group of cells. Here is how you can do those two steps:

1.4.2 Changing the Width of a Column

Objective: To make a column width wider so that all of the information fits inside that column

If you look at your computer screen, you can see that Column B is not wide enough so that all of the information fits inside this column. To make Column B wider:

Click on the letter, B, at the top of your computer screen

Place your mouse pointer at the far right corner of B until you create a "cross sign" on that corner

Left-click on your mouse, hold it down, and move this corner to the right until it is "wide enough to fit all of the data"

Take your finger off the mouse to set the new column width (see Fig. 1.4)

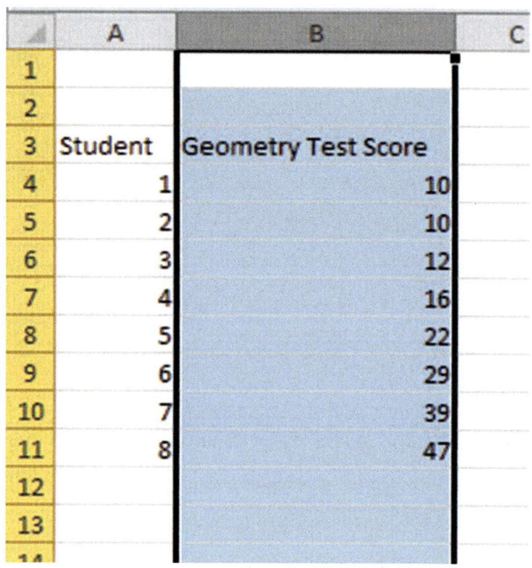

Fig. 1.4 Example of How to Widen the Column Width

Then, click on any empty cell (i.e., any blank cell) to "deselect" column B so that it is no longer a darker color on your screen.

When you widen a column, you will make all of the cells in all of the rows of this column that same width.

Now, let's go through the steps to center the information in both Column A and Column B.

1.4.3 Centering Information in a Range of Cells

Objective: To center the information in a group of cells

In order to make the information in the cells look "more professional," you can center the information using the following steps:

Left-click your mouse on A3 and drag it to the right and down to highlight cells A3:B11 so that these cells appear in a darker color

Home

At the top of your computer screen, you will see a set of "lines" in which all of the lines are "centered" to the same width under "Alignment" (it is the second icon at the bottom left of the Alignment box; see Fig. 1.5)

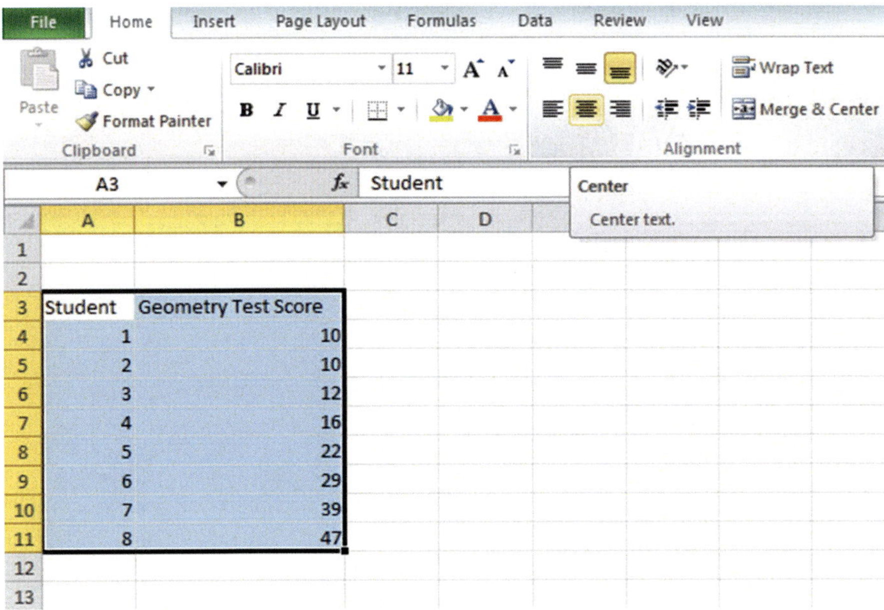

Fig. 1.5 Example of How to Center Information Within Cells

Click on this icon to center the information in the selected cells (see Fig. 1.6)

Fig. 1.6 Final Result of Centering Information in the Cells

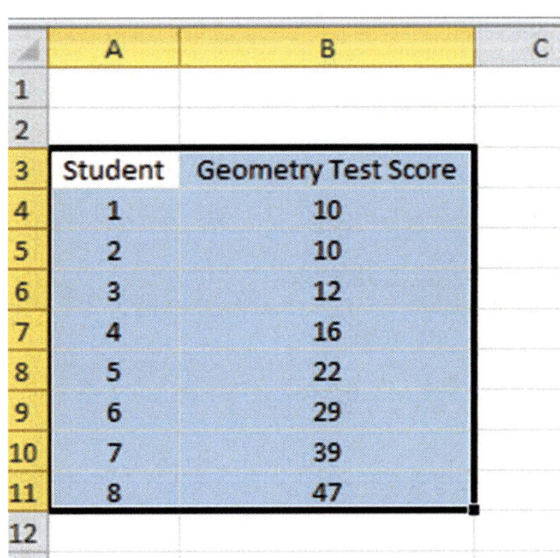

Since you will need to refer to the Geometry Test Scores in your formulas, it will be much easier to do this if you "name the range of data" with a name instead of having to remember the exact cells (B4:B11) in which these figures are located. Let's call that group of cells: Geometry, but we could give them any name that you want to use.

1.4.4 Naming a Range of Cells

Objective: To name the range of data for the test scores with the name: Geometry

Highlight cells B4:B11 by left-clicking your mouse on B4 and dragging it down to B11
Formulas (top left of your screen)
Define Name (top center of your screen)
Geometry (type this name in the top box; see Fig. 1.7)

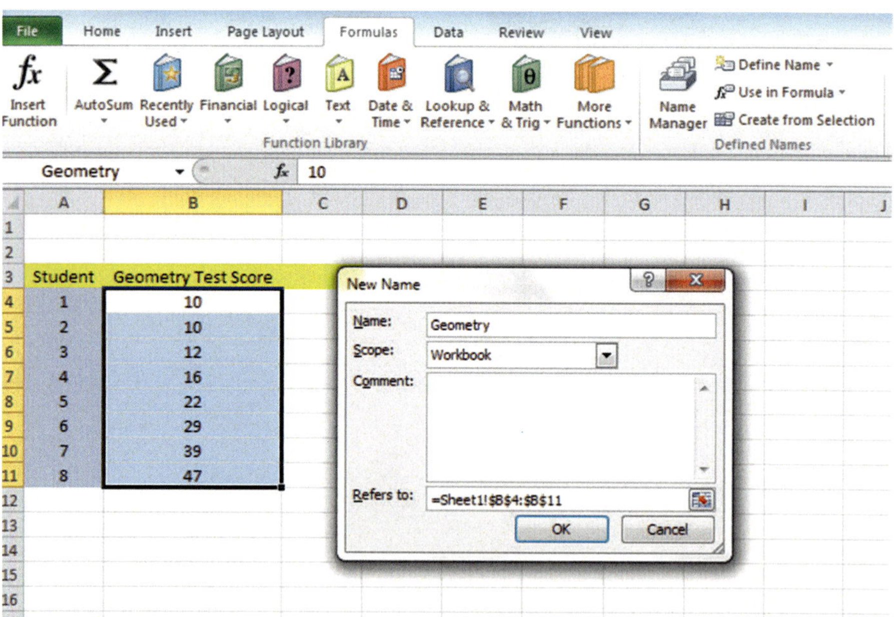

Fig. 1.7 Dialogue box for "naming a range of cells" with the name: Geometry

OK

Then, click on any cell of your spreadsheet that does not have any information in it (i.e., it is an "empty cell") to deselect cells B4:B11

Now, add the following terms to your spreadsheet:

E6: n
E9: Mean
E12: STDEV
E15: s.e. (see Fig. 1.8)

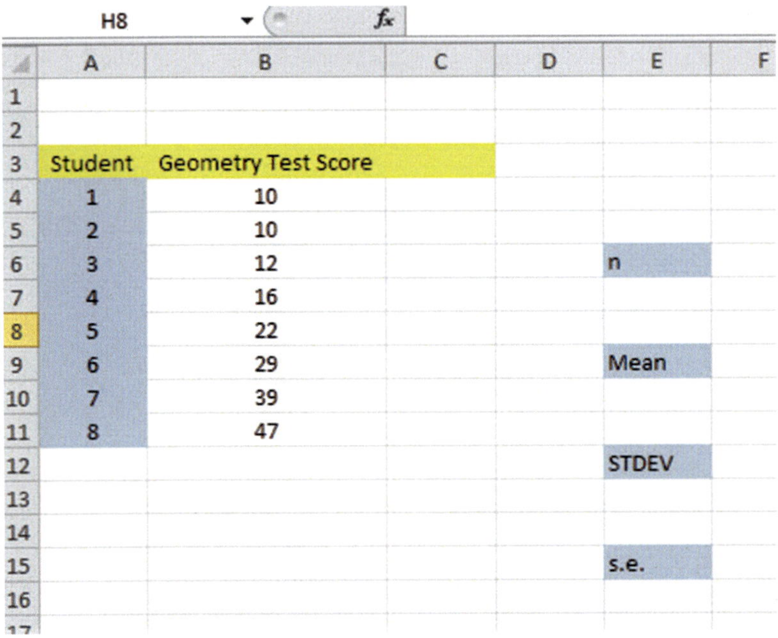

Fig. 1.8 Example of Entering the Sample Size, Mean, STDEV, and s.e. Labels

Note: Whenever you use a formula, you must add an equal sign (=) at the beginning of the name of the function so that Excel knows that you intend to use a formula.

1.4.5 Finding the Sample Size Using the =COUNT Function

Objective: To find the sample size (n) for these data using the =COUNT function

F6: =COUNT(Geometry)

This command should insert the number 8 into cell F6 since there are eight students in this class.

1.4.6 Finding the Mean Score Using the =AVERAGE Function

Objective: To find the mean Geometry figure using the =AVERAGE function

F9: =AVERAGE(Geometry)

This command should insert the number 23.125 into cell F9.

1.4.7 Finding the Standard Deviation Using the =STDEV Function

Objective: To find the standard deviation (STDEV) using the =STDEV function

F12: =STDEV(Geometry)

This command should insert the number 14.02485 into cell F12.

1.4.8 Finding the Standard Error of the Mean

Objective: To find the standard error of the mean using a formula for these eight data points

F15: =F12/SQRT(8)

This command should insert the number 4.958533 into cell F15 (see Fig. 1.9).

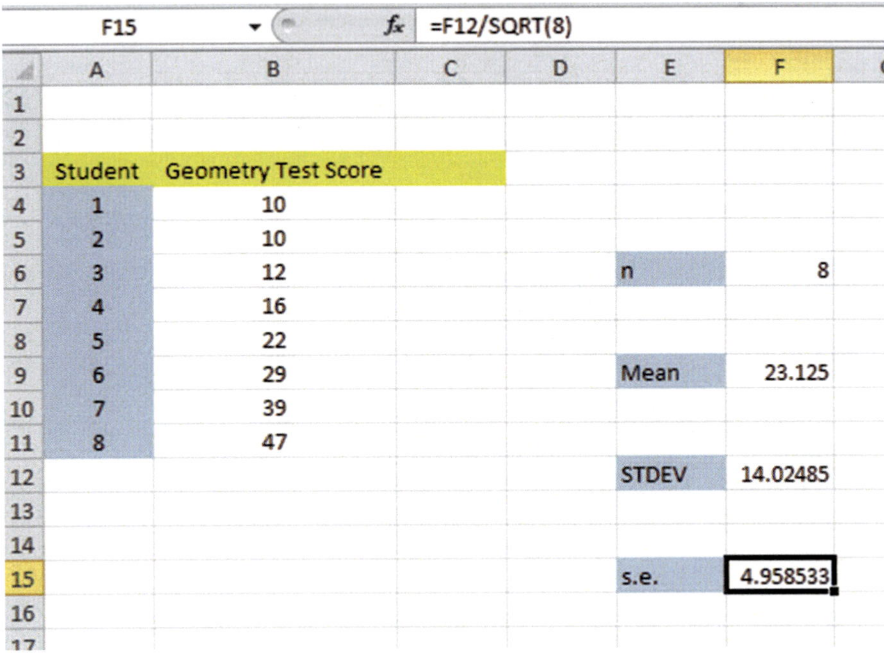

Fig. 1.9 Example of Using Excel Formulas for Sample Size, Mean, STDEV, and s.e.

Important note: Throughout this book, be sure to double-check all of the figures in your spreadsheet to make sure that they are in the correct cells, or the formulas will not work correctly!

1.4.8.1 Formatting Numbers in Number Format (Two Decimal Places)

Objective: To convert the mean, STDEV, and s.e. to two decimal places

Highlight cells F9:F15
Home (top left of screen)
Look under "Number" at the top center of your screen. In the bottom right corner, gently place your mouse pointer on you screen at the bottom of the .00 .0 until it says: "Decrease Decimal" (see Fig. 1.10)

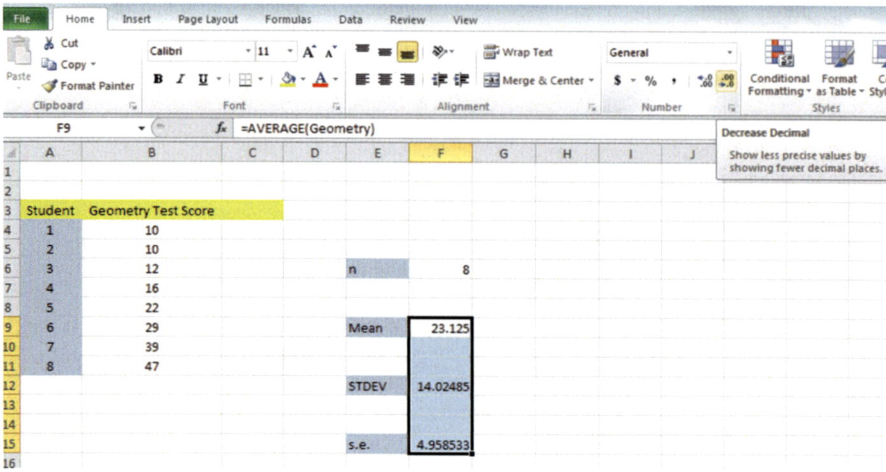

Fig. 1.10 Using the "Decrease Decimal Icon" to convert Numbers to Fewer Decimal Places

Click on this icon *once* and notice that the cells F9:F15 are now all in just two decimal places (see Fig. 1.11)

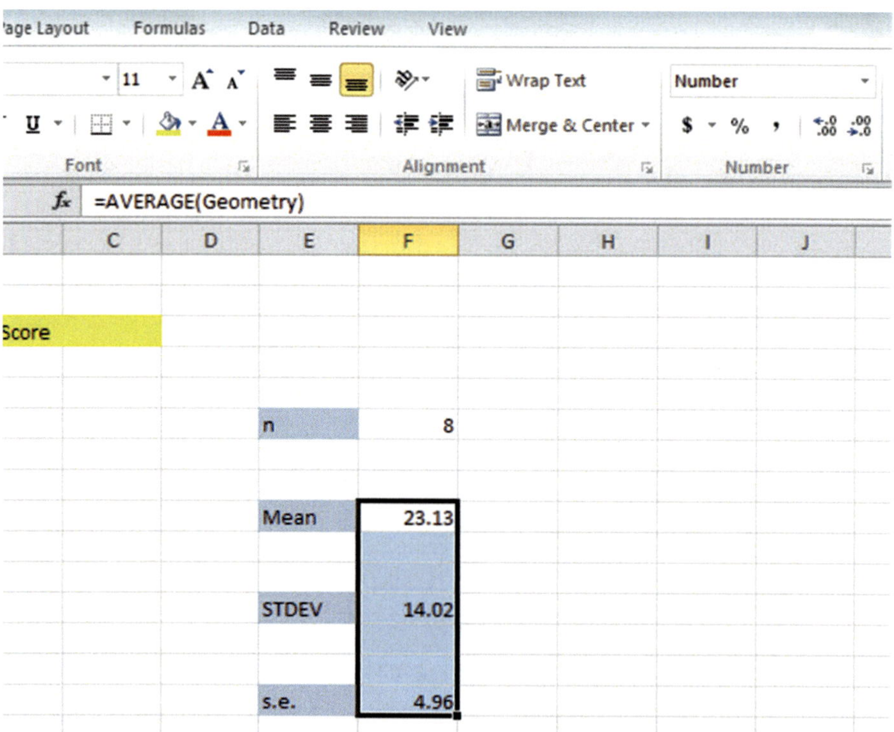

Fig. 1.11 Example of Converting Numbers to Two Decimal Places

Now, click on any "empty cell" on your spreadsheet to deselect cells F9:F15.

1.5 Saving a Spreadsheet

Objective: To save this spreadsheet with the name: Geometry3

In order to save your spreadsheet so that you can retrieve it sometime in the future, your first decision is to decide "where" you want to save it. That is your decision and you have several choices. If it is your own computer, you can save it onto your hard drive (you need to ask someone how to do that on your computer). Or, you can save it onto a "CD" or onto a "flash drive." You then need to complete these steps:

File

Save as

(select the place where you want to save the file by scrolling either down or up the bar on the left, and click on the place where you want to save the file; for example: This PC: My Documents location)

File name: Geometry3 (enter this name to the right of File name; see Fig. 1.12)

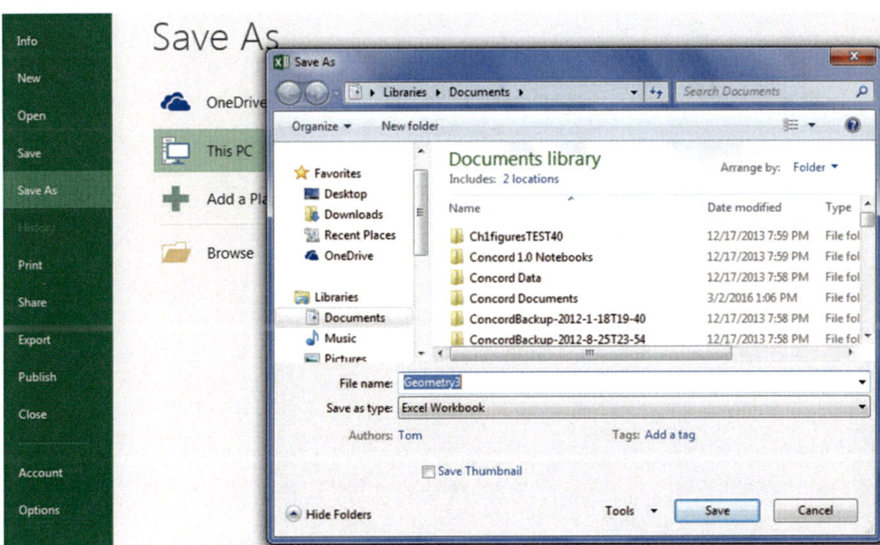

Fig. 1.12 Dialogue Box of Saving an Excel Workbook File as "Geometry3" in My Documents location

Save

Important note: Be very careful to save your Excel file spreadsheet every few minutes so that you do not lose your information!

1.6 Printing a Spreadsheet

Objective: To print the spreadsheet

Use the following procedure when printing any spreadsheet.
File
Print
Print Active Sheets (see Fig. 1.13)

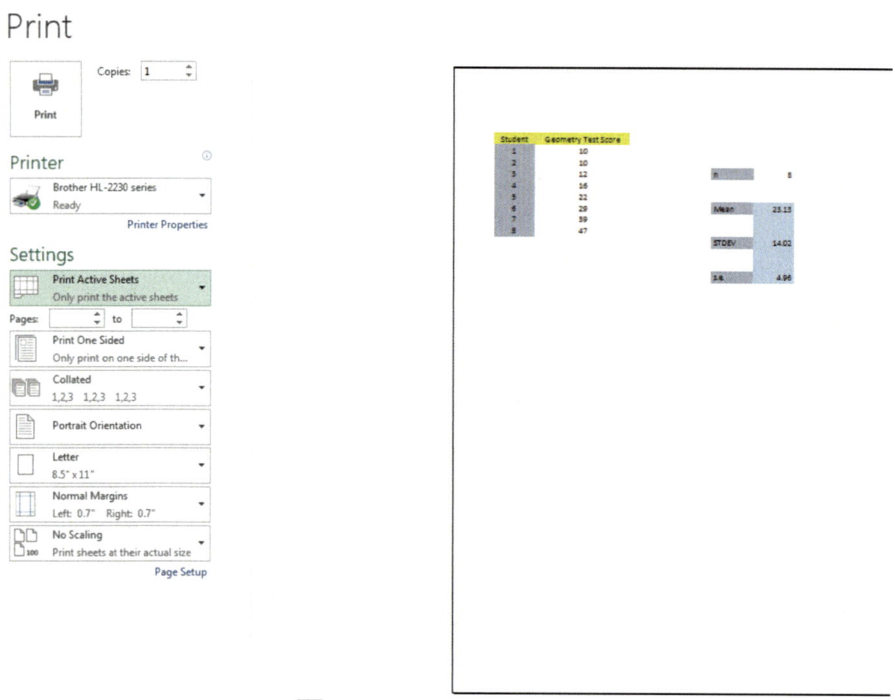

Fig. 1.13 Example of How to Print an Excel Worksheet Using the File/Print/Print Active Sheets Commands

Print (top of your screen)
The final spreadsheet is given in Fig 1.14

Student	Geometry Test Score		
1	10		
2	10		
3	12	n	8
4	16		
5	22		
6	29	Mean	23.13
7	39		
8	47		
		STDEV	14.02
		s.e.	4.96

Fig. 1.14 Final Result of Printing an Excel Spreadsheet

Before you leave this chapter, let's practice changing the format of the figures on a spreadsheet with two examples: (1) using two decimal places for figures that are dollar amounts, and (2) using three decimal places for figures.

Save the final spreadsheet by: File, Save; then close your spreadsheet by: File/Close, and then open a blank Excel spreadsheet by using: File/New/Blank Workbook (in the top left of your screen).

1.7 Formatting Numbers in Currency Format (Two Decimal Places)

Objective: To change the format of figures to dollar format with two decimal places

A3: Price
A4: 1.25
A5: 3.45
A6: 12.95

Highlight cells A4:A6 by left-clicking your mouse on A4 and dragging it down so that these three cells are highlighted in a darker color
Home
Number (top center of screen: click on the down arrow on the right; see Fig. 1.15)

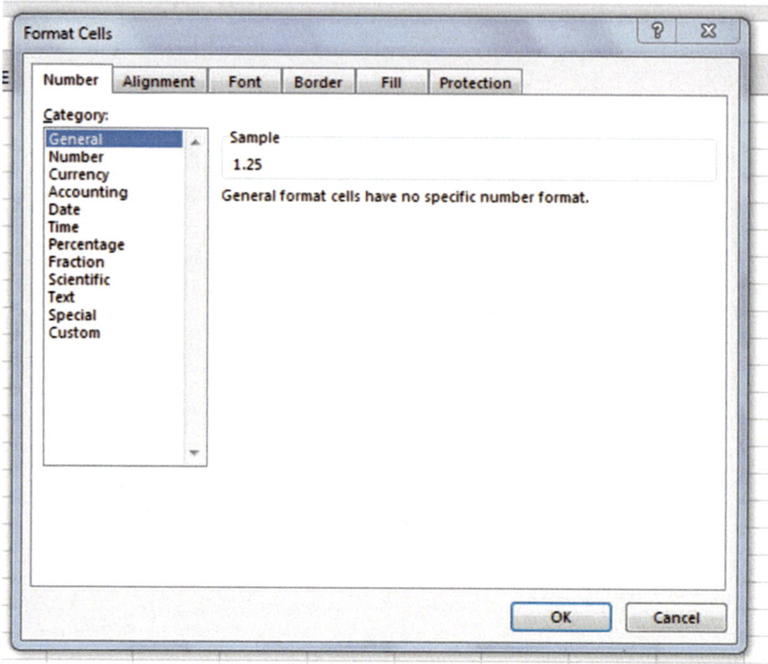

Fig. 1.15 Dialogue Box for Number Format Choices

Category: Currency
Decimal places: 2 (then see Fig. 1.16)

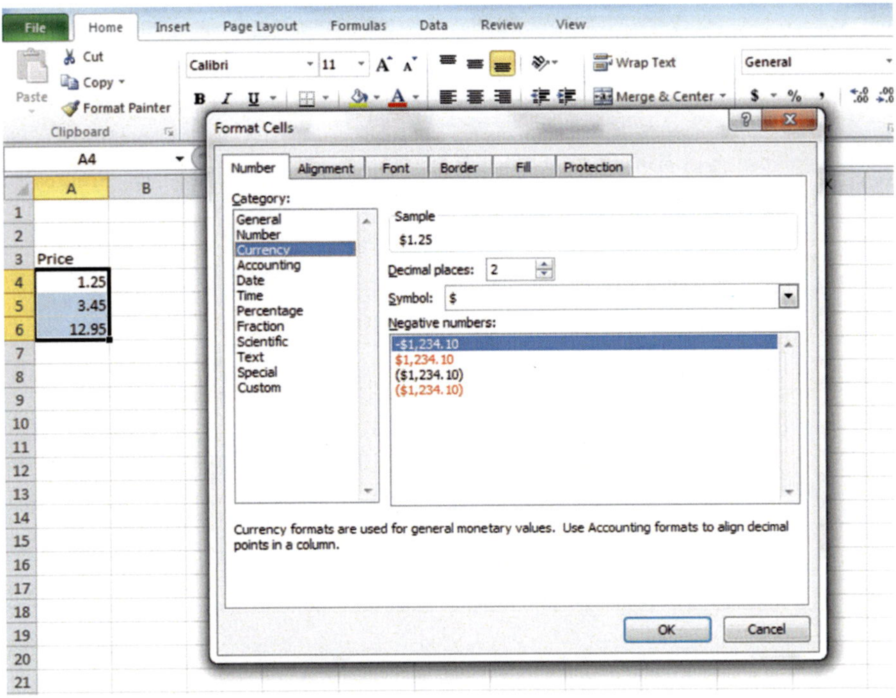

Fig. 1.16 Dialogue Box for Currency (two decimal places) Format for Numbers

OK

The three cells should have a dollar sign in them and be in two decimal places. Next, let's practice formatting figures in number format, three decimal places.

1.8 Formatting Numbers in Number Format (Three Decimal Places)

Objective: To format figures in number format, three decimal places

Home
Highlight cells A4:A6 on your computer screen
Number (click on the down arrow on the right)
Category: number
At the right of the box, change two decimal places to three decimal places by clicking on the "up arrow" once

OK

The three figures should now be in number format, each with three decimals.

Now, click on any blank cell to deselect cells A4:A6. Then, close this file by File/Close/Don't Save (since there is no need to save this practice problem).

You can use these same commands to format a range of cells in percentage format (and many other formats) to whatever number of decimal places you want to specify.

1.9 End-of-Chapter Practice Problems

1. Suppose that a political science professor at a large U.S. university was trying to determine the attitudes of undergraduates at her university regarding U.S.–Chinese relations. She created a survey, and pretests it using a small group of students, and item #8 of this survey resulted in the hypothetical data appearing in Fig. 1.17.

POLITICAL SCIENCE SURVEY OF U.S. - CHINESE RELATIONS

Item #8: "The Chinese leaders are basically trying to get along with the U.S."

1	2	3	4	5	6	7
strongly disagree			undecided			strongly agree

Data
6
4
5
3
2
7
5
3
4
2
4
6
3
5
4
2
1

Fig. 1.17 Worksheet Data for Chap. 1: Practice Problem #1

(a) Use Excel to the right of the table to find the sample size, mean, standard deviation, and standard error of the mean for these data. Label your answers, and round off the mean, standard deviation, and standard error of the mean to two decimal places; use number format for these three figures.
(b) Print the result on a separate page.
(c) Save the file as: China7

2. Suppose that the Human Resources department of your company has administered a "Morale Survey" to all middle-level managers and that you have been asked to summarize the results of the survey. You have decided to test your Excel skills on one item to see if you can do this assignment correctly, and you have selected item #21 to test out your skills. The data are given in Fig. 1.18.

HUMAN RESOURCES MORALE SURVEY

Item #21: **"Management is doing a good job of keeping employee morale at a high level."**

1	2	3	4	5	6	7
Disagree						Agree

Rating
3
6
5
7
2
3
6
5
4
7
6
1
3
2
4
5
6
4
5
3
6
4
7

Fig. 1.18 Worksheet Data for Chap. 1: Practice Problem #2

(a) Use Excel to create a table of these ratings, and at the right of the table use Excel to find the sample size, mean, standard deviation, and standard error of the mean for these data. Label your answers, and round off the mean, standard deviation, and standard error of the mean to two decimal places using number format.
(b) Print the result on a separate page.
(c) Save the file as: MORALE4

3. Suppose that a 5th grade science teacher at Deer Creek Elementary School in Bailey, Colorado, is using a textbook based on basic geology that typically requires about eight class days to teach each chapter. At the end of Chap. 8, the teacher gives a 15-item true-false quiz on this chapter. The test results are given in Fig. 1.19:

Fig. 1.19 Worksheet Data for Chap. 1: Practice Problem #3

Deer Creek Elementary School
5th grade science test
Chapter 8 (15 items)
12
15
13
8
10
12
13
12
9
4
11
15
13
15
12
14

(a) Use Excel to create a table for these data, and at the right of the table, use Excel to find the sample size, mean, standard deviation, and standard error of the mean for these data. Label your answers, and round off the mean, standard deviation, and standard error of the mean to three decimal places using number format.
(b) Print the result on a separate page.
(c) Save the file as: SCIENCE8

References

Neuman, W.L. Social Research Methods: Qualitative and Quantitative Approaches (4th ed.). Boston, MA: Allyn and Bacon, 2000.

Weiers, R.M. Introduction to Business Statistics (7th ed.). Mason, OH: South-Western Cengage Learning, 2011.

Chapter 2
Random Number Generator

Suppose that a local school superintendent asked you to take a random sample of 5 of an elementary school's 32 teachers using Excel so that you could interview these five teachers about their job satisfaction at their school.

To do that, you need to define a "sampling frame." A sampling frame is a list of people from which you want to select a random sample. This frame starts with the identification code (ID) of the number 1 that is assigned to the name of the first teacher in your list of 32 teachers in this school. The second teacher has a code number of 2, the third a code number of 3, and so forth until the last teacher has a code number of 32.

Since this school has 32 teachers, your sampling frame would go from 1 to 32 with each teacher having a unique ID number.

We will first create the frame numbers as follows in a new Excel worksheet:

2.1 Creating Frame Numbers for Generating Random Numbers

Objective: To create the frame numbers for generating random numbers

A3: FRAME NO.
A4: 1

Now, create the frame numbers in column A with the Home/Fill commands that were explained in the first chapter of this book (see Sect. 1.4.1) so that the frame numbers go from 1 to 32, with the number 32 in cell A35. If you need to be reminded about how to do that, here are the steps:

Click on cell A4 to select this cell

© Springer International Publishing Switzerland 2016

T.J. Quirk, *Excel 2016 for Social Science Statistics*, Excel for Statistics,
DOI 10.1007/978-3-319-39711-5_2

Home
Fill (then click on the "down arrow" next to this command and select)
Series (see Fig. 2.1)

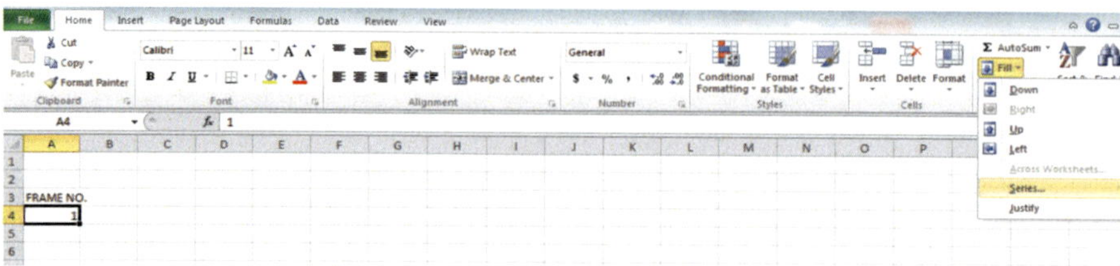

Fig. 2.1 Dialogue Box for Fill/Series Commands

Columns
Step value: 1
Stop value: 32 (see Fig. 2.2)

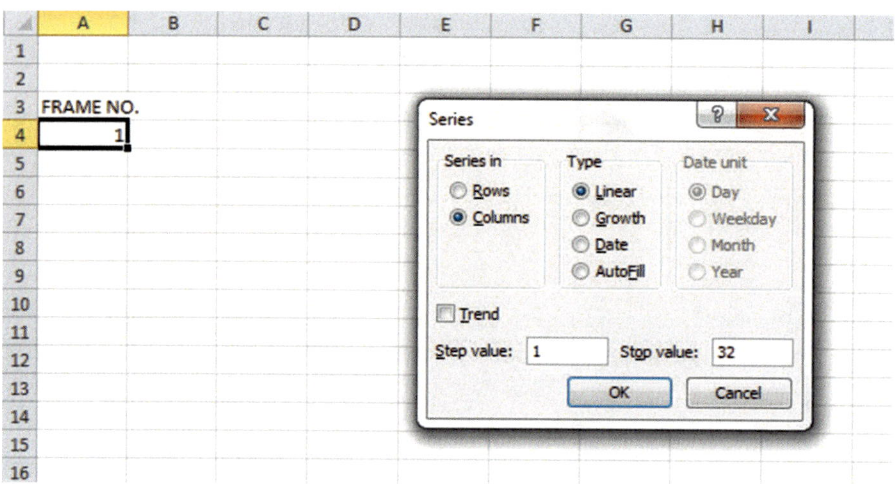

Fig. 2.2 Dialogue Box for Fill/Series/Columns/Step value/ Stop value Commands

OK

Then, save this file as: Random29. You should obtain the result in Fig. 2.3. Now, create a column next to these frame numbers in this manner:

Fig. 2.3 Frame Numbers from 1 to 32

FRAME NO.
1
2
3
4
5
6
7
8
9
10
11
12
13
14
15
16
17
18
19
20
21
22
23
24
25
26
27
28
29
30
31
32

B3: DUPLICATE FRAME NO.
B4: 1

Next, use the Home/Fill command again, so that the 32 frame numbers begin in cell B4 and end in cell B35. Be sure to widen the columns A and B so that all of the information in these columns fits inside the column width. Then, center the information inside both Column A and Column B on your spreadsheet. You should obtain the information given in Fig. 2.4.

Fig. 2.4 Duplicate Frame Numbers from 1 to 32

FRAME NO.	DUPLICATE FRAME NO.
1	1
2	2
3	3
4	4
5	5
6	6
7	7
8	8
9	9
10	10
11	11
12	12
13	13
14	14
15	15
16	16
17	17
18	18
19	19
20	20
21	21
22	22
23	23
24	24
25	25
26	26
27	27
28	28
29	29
30	30
31	31
32	32

Save this file as: Random30

You are probably wondering why you created the same information in both Column A and Column B of your spreadsheet. This is to make sure that before you sort the frame numbers that you have exactly 32 of them when you finish sorting them into a random sequence of 32 numbers.

Now, let's add a random number to each of the duplicate frame numbers as follows:

2.2 Creating Random Numbers in an Excel Worksheet

C3: RANDOM NO.
(then widen columns A, B, C so that their labels fit inside the columns; then center the information in A3:C35)

C4: =RAND()

Next, hit the Enter key to add a random number to cell C4.

Note that you need *both* an open parenthesis *and* a closed parenthesis after =*RAND*(). The RAND command "looks to the left of the cell with the RAND() COMMAND in it" and assigns a random number to that cell.

Now, put the pointer using your mouse in cell C4 and then move the pointer to the bottom right corner of that cell until you see a "plus sign" in that cell. Then, left-click and drag the pointer down to cell C35 to add a random number to all 32 ID frame numbers (see Fig. 2.5).

Fig. 2.5 Example of Random Numbers Assigned to the Duplicate Frame Numbers

FRAME NO.	DUPLICATE FRAME NO.	RANDOM NO.
1	1	0.34366933
2	2	0.209710417
3	3	0.353152217
4	4	0.876383935
5	5	0.122419193
6	6	0.204430049
7	7	0.398194263
8	8	0.324276865
9	9	0.005889939
10	10	0.567422956
11	11	0.142320841
12	12	0.680689895
13	13	0.598004009
14	14	0.681829913
15	15	0.549324011
16	16	0.155400574
17	17	0.897624139
18	18	0.017463156
19	19	0.848841454
20	20	0.037209205
21	21	0.658787315
22	22	0.968460117
23	23	0.275593187
24	24	0.838776061
25	25	0.673063444
26	26	0.281472156
27	27	0.665203225
28	28	0.464583076
29	29	0.314281291
30	30	0.532909472
31	31	0.823737964
32	32	0.956421134

Then, click on any empty cell to deselect C4:C35 to remove the dark color highlighting these cells.

Save this file as: Random31

Now, let's sort these duplicate frame numbers into a random sequence:

2.3 Sorting Frame Numbers into a Random Sequence

Objective: To sort the duplicate frame numbers into a random sequence

Highlight cells B3:C35 (include the labels at the top of columns B and C)
Data (top of screen)
Sort (click on this word at the top center of your screen; see Fig. 2.6)

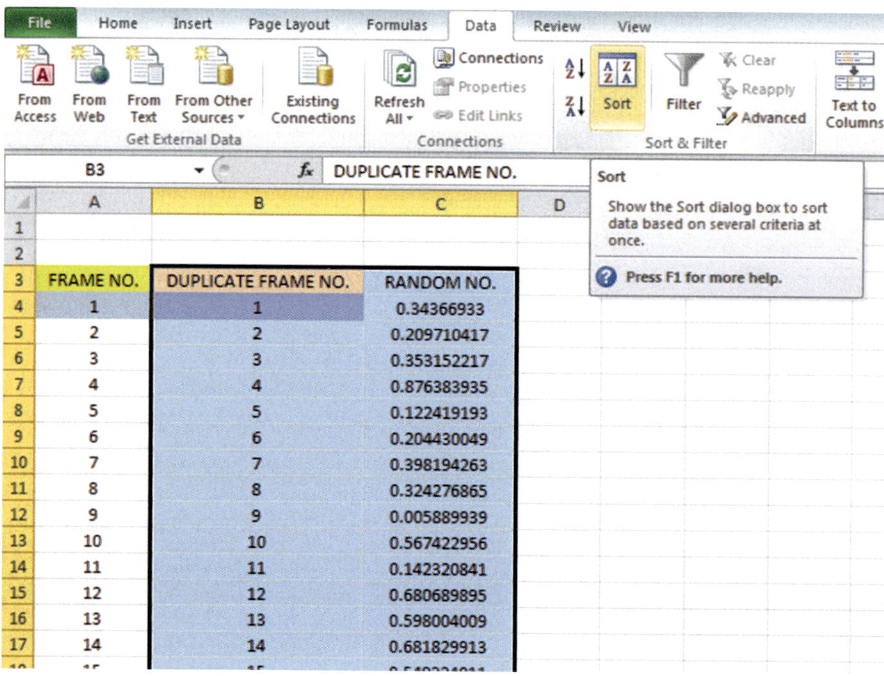

Fig. 2.6 Dialogue Box for Data/Sort Commands

Sort by: RANDOM NO. (click on the down arrow)
Smallest to Largest (see Fig. 2.7)

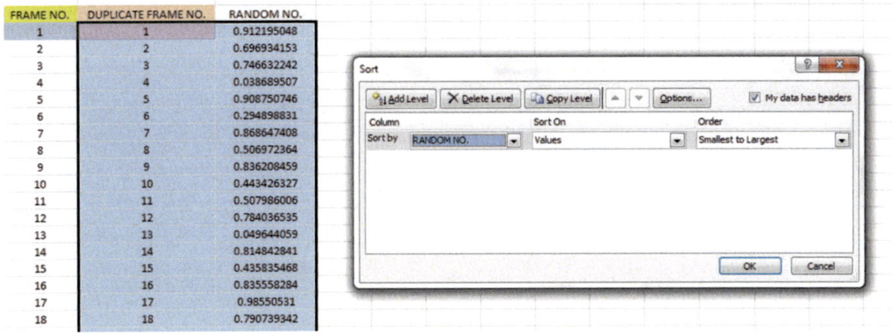

Fig. 2.7 Dialogue Box for Data/Sort/RANDOM NO./Smallest to Largest Commands

OK
Click on any empty cell to deselect B3:C35.
Save this file as: Random32
Print this file now.

These steps will produce Fig. 2.8 with the DUPLICATE FRAME NUMBERS sorted into a random order:

Important note: Because Excel randomly assigns these random numbers, your Excel commands will produce a different sequence of random numbers from everyone else who reads this book!

Fig. 2.8 Duplicate Frame Numbers Sorted by Random Number

FRAME NO.	DUPLICATE FRAME NO.	RANDOM NO.
1	9	0.079359184
2	5	0.063853375
3	14	0.010194229
4	1	0.395726092
5	10	0.272449497
6	13	0.751154385
7	8	0.374257307
8	28	0.931592868
9	11	0.784319855
10	18	0.062115733
11	32	0.010004276
12	21	0.121391862
13	2	0.04094519
14	26	0.045169691
15	24	0.019558689
16	31	0.654079136
17	25	0.218618972
18	27	0.846341916
19	23	0.461139619
20	20	0.040189725
21	12	0.561605359
22	22	0.660369898
23	19	0.385059983
24	3	0.141596898
25	30	0.884990632
26	29	0.478518
27	4	0.049860437
28	15	0.644759216
29	7	0.749186612
30	6	0.304230163
31	17	0.717396227
32	16	0.256921702

Because your objective at the beginning of this chapter was to select randomly 5 of this school's 32 teachers for a personal interview, you now can do that by selecting the *first five ID numbers* in DUPLICATE FRAME NO. column after the sort.

Although your first five random numbers will be different from those we have selected in the random sort that we did in this chapter, we would select these five IDs of teachers to interview using Fig. 2.9.

Fig. 2.9 First Five Teachers Selected Randomly

FRAME NO.	DUPLICATE FRAME NO.	RANDOM NO.
1	9	0.079359184
2	5	0.063853375
3	14	0.010194229
4	1	0.395726092
5	10	0.272449497
6	13	0.751154385
7	8	0.374257307
8	28	0.931592868
9	11	0.784319855
10	18	0.062115733
11	32	0.010004276
12	21	0.121391862
13	2	0.04094519
14	26	0.045169691
15	24	0.019558689
16	31	0.654079136
17	25	0.218618972
18	27	0.846341916
19	23	0.461139619
20	20	0.040189725
21	12	0.561605359
22	22	0.660369898
23	19	0.385059983
24	3	0.141596898
25	30	0.884990632
26	29	0.478518
27	4	0.049860437
28	15	0.644759216
29	7	0.749186612
30	6	0.304230163
31	17	0.717396227
32	16	0.256921702

9, 5, 14, 1, 10

Save this file as: Random33

Remember, your five ID numbers selected after your random sort will be different from the five ID numbers in Fig. 2.9 because Excel assigns a different random number *each time the =RAND() command is given.*

If you want to learn more about the purpose of taking a random sample in social science research, see Frankfort-Nachmias and Nachmias (2008).

Before we leave this chapter, you need to learn how to print a file so that all of the information on that file fits onto a single page without "dribbling over" onto a second or third page.

2.4 Printing an Excel File So That All of the Information Fits onto One Page

Objective: To print a file so that all of the information fits onto one page

Note that the three practice problems at the end of this chapter require you to sort random numbers when the files contain 63 children, 114 counties of the state of Missouri, and 76 key accounts, respectively. These files will be "too big" to fit onto one page when you print them unless you format these files so that they fit onto a single page when you print them.

Let's create a situation where the file does not fit onto one printed page unless you format it first to do that.

Go back to the file you just created, Random 33, and enter the name: *Jennifer* into cell: A50.

If you printed this file now, the name, *Jennifer*, would be printed onto a second page because it "dribbles over" outside of the page rage for this file in its current format.

So, you would need to change the page format so that all of the information, including the name, Jennifer, fits onto just one page when you print this file by using the following steps:

Page Layout (top left of the computer screen)
(Notice the "Scale to Fit" section in the center of your screen; see Fig. 2.10)

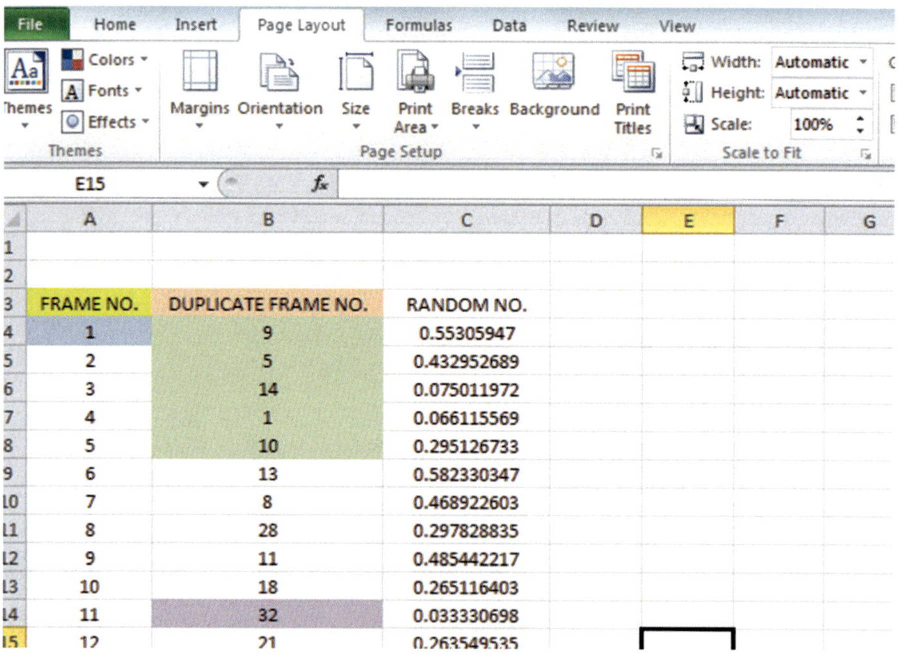

Fig. 2.10 Dialogue Box for Page Layout/Scale to Fit Commands

Hit the down arrow to the right of 100 % *once* to reduce the size of the page to 95 %

Now, note that the name, Jennifer, is still on a second page on your screen because her name is below the horizontal dotted line on your screen in Fig. 2.11 (the dotted lines tell you outline dimensions of the file if you printed it now).

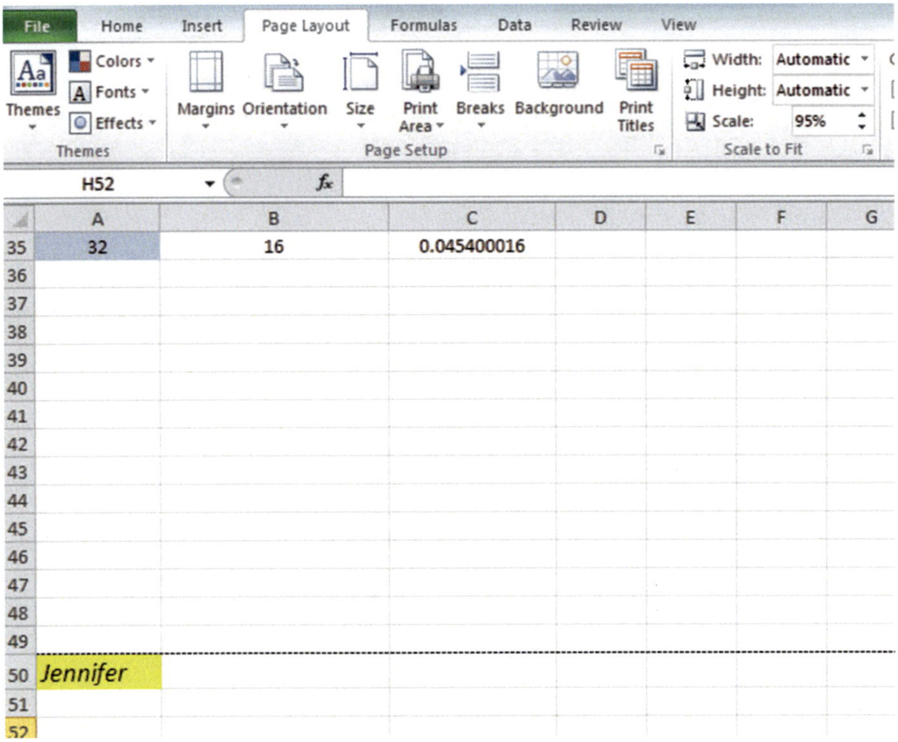

Fig. 2.11 Example of Scale Reduced to 95 % with "Jennifer" to be Printed on a Second Page

So, you need to repeat the "scale change steps" by hitting the down arrow on the right once more to reduce the size of the worksheet to 90 % of its normal size.

Notice that the "dotted lines" on your computer screen in Fig. 2.12 are now below Jennifer's name to indicate that all of the information, including her name, is now formatted to fit onto just one page when you print this file.

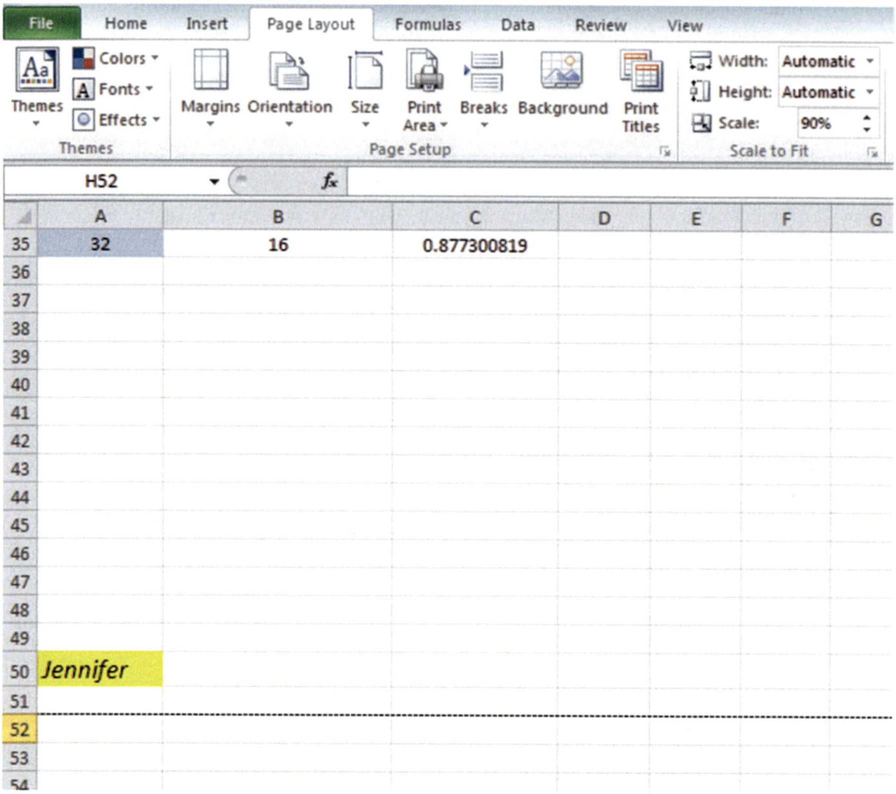

Fig. 2.12 Example of Scale Reduced to 90 % with "Jennifer" to be printed on the first page (note the *dotted line* below Jennifer on your screen)

Save the file as: Random46

Print the file. Does it all fit onto one page? It should (see Fig. 2.13).

Fig. 2.13 Final Spreadsheet of 90 % Scale to Fit

FRAME NO.	DUPLICATE FRAME NO.	RANDOM NO.
1	9	0.101054292
2	5	0.316815602
3	14	0.193676355
4	1	0.10588154
5	10	0.451080782
6	13	0.355938622
7	8	0.068021169
8	28	0.784501566
9	11	0.585160127
10	18	0.700117635
11	32	0.915540411
12	21	0.221331619
13	2	0.711309403
14	26	0.440254429
15	24	0.655728526
16	31	0.10816893
17	25	0.829399085
18	27	0.590317067
19	23	0.115637758
20	20	0.515528511
21	12	0.739255415
22	22	0.854646809
23	19	0.798190748
24	3	0.515150748
25	30	0.279981538
26	29	0.331013093
27	4	0.566738676
28	15	0.118570924
29	7	0.485341159
30	6	0.423541413
31	17	0.507712141
32	16	0.877300819
Jennifer		

2.5 End-of-Chapter Practice Problems

1. Howell et al. (2000) studied the effects of a private school education on poor children. Programs were implemented in three U.S. cities that offered poor children in public schools partial scholarship vouchers to attend private schools.

Since more children applied for the program than could be accepted into the program, children were selected randomly to participate in the voucher program. Suppose that you were assigned the task of selecting the children randomly for this program, and that you wanted to test your Excel skills by selecting 15 of 63 children who applied for this program before you did the actual random selection of all of the students who applied for the program.

(a) Set up a spreadsheet of frame numbers for these children with the heading: FRAME NUMBERS using the Home/Fill commands.
(b) Then, create a separate column to the right of these frame numbers which duplicates these frame numbers with the title: Duplicate frame numbers
(c) Then, create a separate column to the right of these duplicate frame numbers and use the =RAND() function to assign random numbers to all of the frame numbers in the duplicate frame numbers column, and change this column format so that three decimal places appear for each random number
(d) Sort the duplicate frame numbers and random numbers into a random order
(e) Print the result so that the spreadsheet fits onto one page
(f) Circle on your printout the I.D. number of the first 15 children that you would call in your phone survey
(g) Save the file as: RAND9

Important note: Note that everyone who does this problem will generate a different random order of children ID numbers since Excel assigns a different random number each time the RAND() command is used. For this reason, the answer to this problem given in this Excel Guide will have a completely different sequence of random numbers from the random sequence that you generate. This is normal and what is to be expected.

2. Suppose that you wanted to do a random sample of 10 of the 114 counties in the state of Missouri as requested by a political pollster who wants to select registered voters by county in Missouri for a phone survey of their voting preferences in the next election. You know that there are 114 counties in Missouri because you have accessed the Web site for the U.S. census (U.S. Census Bureau 2000). For your information, the United States has a total of 3,140 counties in its 50 states (U.S. Census Bureau 2000).

(a) Set up a spreadsheet of frame numbers for these counties with the heading: FRAME NO.
(b) Then, create a separate column to the right of these frame numbers which duplicates these frame numbers with the title: Duplicate frame no.
(c) Then, create a separate column to the right of these duplicate frame numbers entitled "Random number" and use the =RAND() function to assign random numbers to all of the frame numbers in the duplicate frame numbers column. Then, change this column format so that three decimal places appear for each random number
(d) Sort the duplicate frame numbers and random numbers into a random order

(e) Print the result so that the spreadsheet fits onto one page
(f) Circle on your printout the I.D. number of the first 10 counties that the pollster would call in his phone survey
(g) Save the file as: RANDOM6

3. Suppose that a Sales department at a company wants to do a "customer satisfaction survey" of 20 of this company's 76 "key accounts." Suppose, further, that the Sales Vice-President has defined a key account as a customer who purchased at least $30,000 worth of merchandise from this company in the past 90 days.

   (a) Set up a spreadsheet of frame numbers for these customers with the heading: FRAME NUMBERS.
   (b) Then, create a separate column to the right of these frame numbers which duplicates these frame numbers with the title: Duplicate frame numbers
   (c) Then, create a separate column to the right of these duplicate frame numbers entitled "Random number" and use the =RAND() function to assign random numbers to all of the frame numbers in the duplicate frame numbers column. Then, change this column format so that three decimal places appear for each random number
   (d) Sort the duplicate frame numbers and random numbers into a random order
   (e) Print the result so that the spreadsheet fits onto one page
   (f) Circle on your printout the I.D. number of the first 20 customers that the Sales Vice-President would call for his phone survey.
   (g) Save the file as: RAND5

References

Frankfort-Nachmias, C. and Nachmias, D. Research Methods in the Social Sciences (7th ed.). New York, NY: Worth Publishers, 2008.

Howell, W.G., Wolf, P.J., Campbell, D.E., and Peterson, P.E. "Test-Score Effects of School Vouchers in Dayton, Ohio, New York City, and Washington, D.C.: Evidence from Randomized Field Trials." Paper presented at the annual meeting of the American Political Science Association, Washington, D.C., September 2000.

U.S. Census Bureau Census 2000 PHC-T-4. Ranking tables for counties 1990 and 2000. Retrieved from http://www.census.gov/population/www/cen2000/briefs/phc-t4/tables/tab01.pdf

Chapter 3
Confidence Interval About the Mean Using the TINV Function and Hypothesis Testing

This chapter focuses on two ideas: (1) finding the 95 % confidence interval about the mean, and (2) hypothesis testing.

Let's talk about the confidence interval first.

3.1 Confidence Interval About the Mean

In statistics, we are always interested in *estimating the population mean*. How do we do that?

3.1.1 How to Estimate the Population Mean

Objective: To estimate the population mean, μ

Remember that the population mean is the average of all of the people in the target population. For example, if we were interested in how well adults ages 25–44 liked a new flavor of Ben & Jerry's ice cream, we could never ask this question of all of the people in the U.S. who were in that age group. Such a research study would take way too much time to complete and the cost of doing that study would be prohibitive.

So, instead of testing *everyone* in the population, we take a sample of people in the population and use the results of this sample to estimate the mean of the entire population. This saves both time and money. When we use the results of a sample to estimate the population mean, this is called "*inferential statistics*" because we are inferring the population mean from the sample mean (see e.g. King et al. 1994).

© Springer International Publishing Switzerland 2016

T.J. Quirk, *Excel 2016 for Social Science Statistics*, Excel for Statistics,
DOI 10.1007/978-3-319-39711-5_3

When we study a sample of people in social science research, we know the size of our sample (n), the mean of our sample ($\overline{X}$), and the standard deviation of our sample (STDEV). We use these figures to estimate the population mean with a test called the "confidence interval about the mean."

3.1.2 Estimating the Lower Limit and the Upper Limit of the 95 Percent Confidence Interval About the Mean

The theoretical background of this test is beyond the scope of this book, and you can learn more about this test from studying any good statistics textbook (e.g. Levine 2011; Pollock 2009) but the basic ideas are as follows.

We assume that the population mean is somewhere in an interval which has a "lower limit" and an "upper limit" to it. We also assume in this book that we want to be "95 % confident" that the population mean is inside this interval somewhere. So, we intend to make the following type of statement:

> "We are 95 % confident that the population mean in miles per gallon (mpg) for the Chevy Impala automobile is between 26.92 miles per gallon and 29.42 miles per gallon."

If we want to create a billboard for this car that claims that this car gets 28 miles per gallon (mpg), we can do that because 28 is *inside the 95 % confidence interval* in our research study in the above example. We do not know exactly what the population mean is, only that it is somewhere between 26.92 mpg and 29.42 mpg, and 28 is inside this interval.

But we are only 95 % confident that the population mean is inside this interval, and 5 % of the time we will be wrong in assuming that the population mean is 28 mpg.

But, for our purposes in social science research, we are happy to be 95 % confident that our assumption is accurate. We should also point out that 95 % is an arbitrary level of confidence for our results. We could choose to be 80 % confident, or 90 % confident, or even 99 % confident in our results if we wanted to do that. But, in this book, *we will always assume that we want to be 95 % confident of our results*. That way, you will not have to guess on how confident you want to be in any of the problems in this book. We will always want to be 95 % confident of our results in this book.

So how do we find the 95 % confidence interval about the mean for our data?

In words, we will find this interval this way:

> "Take the sample mean ($\overline{X}$), *and add to it* 1.96 times the standard error of the mean (s.e.) to get the upper limit of the confidence interval. Then, take the sample mean, *and subtract from it* 1.96 times the standard error of the mean to get the lower limit of the confidence interval."

You will remember (see Sect. 1.3) that the standard error of the mean (s.e.) is found by dividing the standard deviation of our sample (STDEV) by the square root of our sample size, n.

In mathematical terms, the formula for the 95 % confidence interval about the mean is:

$$\overline{X} \pm 1.96 \text{ s.e.} \tag{3.1}$$

Note that the"± *sign*" stands for "plus or minus," and this means that you first add 1.96 times the s.e. to the mean to get the upper limit of the confidence interval, and then subtract 1.96 times the s.e. from the mean to get the lower limit of the confidence interval. Also, the symbol 1.96 s.e. means that you multiply 1.96 times the standard error of the mean to get this part of the formula for the confidence interval.

Note: We will explain shortly where the number 1.96 came from

Let's try a simple example to illustrate this formula.

3.1.3 Estimating the Confidence Interval for the Chevy Impala in Miles per Gallon

Let's suppose that you asked owners of the Chevy Impala to keep track of their mileage and the number of gallons used for two tanks of gas. Let's suppose that 49 owners did this, and that they average 27.83 miles per gallon (mpg) with a standard deviation of 3.01 mpg. The standard error (s.e.) would be 3.01 divided by the square root of 49 (i.e., 7) which gives a s.e. equal to 0.43.

The 95 % confidence interval for these data would be:

$$27.83 \pm 1.96(0.43)$$

The *upper limit of this confidence interval* uses the plus sign of the ± sign in the formula. Therefore, the upper limit would be:

$$27.83 + 1.96(0.43) = 27.83 + 0.84 = 28.67 \text{mpg}$$

Similarly, *the lower limit of this confidence interval* uses the minus sign of the ± sign in the formula. Therefore, the lower limit would be:

$$27.83 - 1.96(0.43) = 27.83 - 0.84 = 26.99 \text{mpg}$$

The result of our research study would, therefore, be the following:

> "We are 95 % confident that the population mean for the Chevy Impala is somewhere between 26.99 mpg and 28.67 mpg."

If we were planning to create a billboard that claimed that this car got 28 mpg, we would be able to do that based on our data, since 28 is inside of this 95 % confidence interval for the population mean.

You are probably asking yourself: "Where did that 1.96 in the formula come from?"

3.1.4 Where Did the Number "1.96" Come From?

A detailed mathematical answer to that question is beyond the scope of this book, but here is the basic idea.

We make an assumption that the data in the population are "normally distributed" in the sense that the population data would take the shape of a "normal curve" if we could test all of the people in the population. The normal curve looks like the outline of the Liberty Bell that sits in front of Independence Hall in Philadelphia, Pennsylvania. The normal curve is "symmetric" in the sense that if we cut it down the middle, and folded it over to one side, the half that we folded over would fit perfectly onto the half on the other side. If you want to learn more about the normal curve, see Steinberg (2008) and Frankfort-Nachmias and Nachmias (2008).

A discussion of integral calculus is beyond the scope of this book, but essentially we want to find the lower limit and the upper limit of the population data in the normal curve so that 95 % of the area under this curve is between these two limits. *If we have more than 40 people in our research study*, the value of these limits is plus or minus 1.96 times the standard error of the mean (s.e.) of our sample. The number 1.96 times the s.e. of our sample gives us the upper limit and the lower limit of our confidence interval. If you want to learn more about this idea, you can consult a good statistics book (e.g. Salkind 2010).

The number 1.96 would change if we wanted to be confident of our results at a different level from 95 % as long as we have more than 40 people in our research study.

For example:

1. If we wanted to be 80 % confident of our results, this number would be 1.282.
2. If we wanted to be 90 % confident of our results, this number would be 1.645.
3. If we wanted to be 99 % confident of our results, this number would be 2.576.

But since we always want to be 95 % confident of our results in this book, we will always use 1.96 in this book whenever we have more than 40 people in our research study.

By now, you are probably asking yourself: "Is this number in the confidence interval about the mean always 1.96?" The answer is: "No!", and we will explain why this is true now.

3.1.5 Finding the Value for t in the Confidence Interval Formula

Objective: To find the value for t in the confidence interval formula

The correct formula for the confidence interval about the mean for different sample sizes is the following:

$$\overline{X} \pm \text{t s.e.} \tag{3.2}$$

To use this formula, you find the sample mean, $\overline{X}$, *and add to it the value of t times the s.e. to get the upper limit* of this 95 % confidence interval. Also, you take the sample mean, $\overline{X}$, and *subtract from it the value of t times the s.e. to get the lower limit* of this 95 % confidence interval. And, you find the value of t in the table given in Appendix E of this book in the following way:

Objective: To find the value of t in the t-table in Appendix E

Before we get into an explanation of what is meant by "the value of t," let's give you practice in finding the value of t by using the t-table in Appendix E.

Keep your finger on Appendix E as we explain how you need to "read" that table.

Since the test in this chapter is called the "confidence interval about the mean test," you will use the first column on the left in Appendix E to find the critical value of t for your research study (note that this column is headed: "sample size n").

To find the value of t, you go down this first column until you find the sample size in your research study, and then you go to the right and read the value of t for that sample size in the "critical t column" of the table (note that this column is the column that you would use for the 95 % confidence interval about the mean).

For example, if you have 14 people in your research study, the value of t is 2.160.

If you have 26 people in your research study, the value of t is 2.060.

If you have more than 40 people in your research study, the value of t is always 1.96.

Note that the "critical t column" in Appendix E represents the value of t that you need to use to obtain to be 95 % confident of your results as "significant" results.

Throughout this book, we are assuming that you want to be 95 % confident in the results of your statistical tests. Therefore, the value for t in the t-table in Appendix E tells you which value you should use for t when you use the formula for the 95 % confidence interval about the mean.

Now that you know how to find the value of t in the formula for the confidence interval about the mean, let's explore how you find this confidence interval using Excel.

3.1.6 *Using Excel's TINV Function to Find the Confidence Interval About the Mean*

Objective: To use the TINV function in Excel to find the confidence interval about the mean

When you use Excel, the formulas for finding the confidence interval are:

Lower limit: $= \overline{X} - TINV(1 - 0.95, n - 1)*s.e.$ (no spaces between these symbols) (3.3)

Upper limit: $= \overline{X} + TINV(1 - 0.95, n - 1)*s.e.$ (no spaces between these symbols) (3.4)

Note that the "**symbol*" in this formula tells Excel to use the multiplication step in the formula, and it stands for "times" in the way we talk about multiplication.

You will recall from Chap. 1 that n stands for the sample size, and so $n - 1$ stands for the sample size minus one.

You will also recall from Chap. 1 that the standard error of the mean, s.e., equals the STDEV divided by the square root of the sample size, n (see Sect. 1.3).

Let's try a sample problem using Excel to find the 95 % confidence interval about the mean for a problem.

Let's suppose that General Motors wanted to claim that its Chevy Impala achieves 28 miles per gallon (mpg). Let's call 28 mpg the "reference value" for this car.

Suppose that you work for Ford Motor Co. and that you want to check this claim to see is it holds up based on some research evidence. You decide to collect some data and to use a two-side 95 % confidence interval about the mean to test your results:

3.1.7 *Using Excel to Find the 95 Percent Confidence Interval for a Car's mpg Claim*

Objective: To analyze the data using a two-side 95 % confidence interval about the mean

You select a sample of 25 new car owners for this car and they agree to keep track of their mileage for two tanks of gas and to record the average miles per gallon they achieve on these two tanks of gas. Your research study produces the results given in Fig. 3.1:

Chevy Impala
Miles per gallon
30.9
24.5
31.2
28.7
35.1
29.0
28.8
23.1
31.0
30.2
28.4
29.3
24.2
27.0
26.7
31.0
23.5
29.4
26.3
27.5
28.2
28.4
29.1
21.9
30.9

Fig. 3.1 Worksheet Data for Chevy Impala (Practical Example)

Create a spreadsheet with these data and use Excel to find the sample size (n), the mean, the standard deviation (STDEV), and the standard error of the mean (s.e.) for these data using the following cell references.

A3: Chevy Impala
A5: Miles per gallon
A6: 30.9

Enter the other mpg data in cells A7:A30

Now, highlight cells A6:A30 and format these numbers in number format (one decimal place). Center these numbers in Column A. Then, widen columns A and B by making both of them twice as wide as the original width of column A. Then, widen column C so that it is three times as wide as the original width of column A so that your tablelooks more professional.

C7: n
C10: Mean
C13: STDEV
C16: s.e.

C19: 95 % confidence interval
D21: Lower limit:
D23: Upper limit: (see Fig. 3.2)

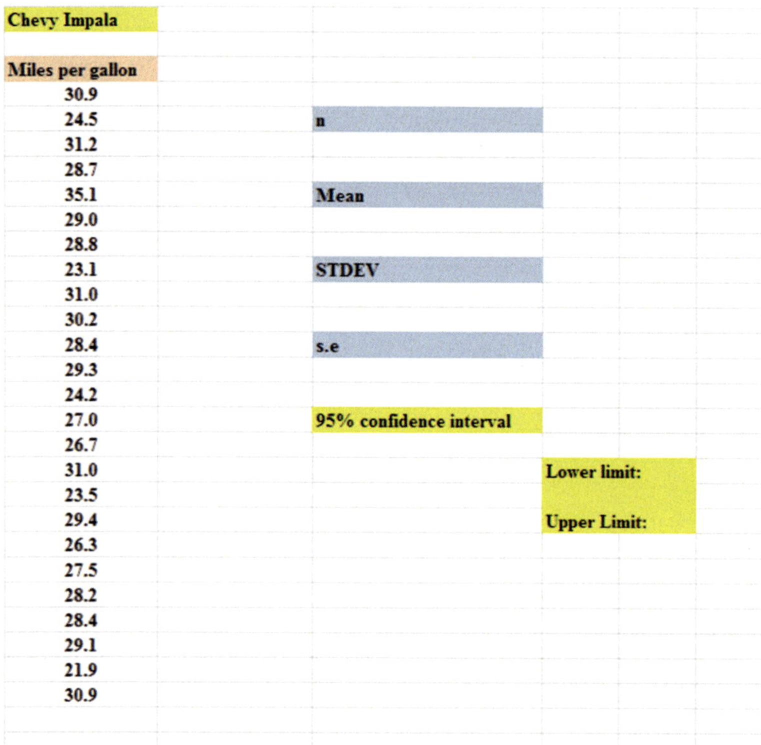

Fig. 3.2 Example of Chevy Impala Format for the Confidence Interval About the Mean Labels

B26: Draw a picture below this confidence interval
B28: ‘26.92 --------------- (be sure to add the single quotation mark before 26.92 so that Excel treats this as a label, instead of a number)

B29: lower
B30: limit
C28: ‘-------- 28 -------28.17 ---------- (note that you need to begin cell C28 with a *single quotation mark* (‘) to tell Excel that this is a *label*, and not a number)
D28: ‘------------ (be sure to use a single quotation mark at the beginning)
E28: ‘29.42 (note the single quotation mark)
C29: ref. Mean

C30: value
E29: upper
E30: limit
B33: Conclusion:

Now, align the labels underneath the picture of the confidence interval so that they look like Figure 3.3.

Chevy Impala					
Miles per gallon					
30.9					
24.5		n			
31.2					
28.7					
35.1		Mean			
29.0					
28.8					
23.1		STDEV			
31.0					
30.2					
28.4		s.e			
29.3					
24.2					
27.0		95% confidence interval			
26.7					
31.0			Lower limit:		
23.5					
29.4			Upper Limit:		
26.3					
27.5					
28.2	Draw a picture below this confidence interval				
28.4					
29.1	26.92 ---------	28 --------- 28.17 ---------------		29.42	
21.9	lower	ref. Mean		upper	
30.9	limit	value		limit	
	Conclusion:				

Fig. 3.3 Example of Drawing a Picture of a Confidence Interval About the Mean Result

Next, name the range of data from A6:A30 as: miles

D7: Use Excel to find the sample size
D10: Use Excel to find the mean
D13: Use Excel to find the STDEV
D16: Use Excel to find the s.e.

Now, you need to find the lower limit and the upper limit of the 95 % confidence interval for this study.

We will use Excel's TINV function to do this. We will assume that you want to be 95 % confident of your results.

$$\text{F21}: \quad =\text{D10} - \text{TINV}(1-.95,24)^{*}\text{D16} \quad \text{(no spaces between symbols)}$$

Note that this TINV formula uses 24 since 24 is one less than the sample size of 25 (i.e., 24 is n − 1). Note that D10 is the mean, while D16 is the standard error of the mean. The above formula gives the *lower limit of the confidence interval, 26.92.*

$$\text{F23}: \quad =\text{D10} + \text{TINV}(1-.95,24)^{*}\text{D16} \quad \text{(no spaces between symbols)}$$

The above formula gives the *upper limit of the confidence interval, 29.42.*

Now, use number format (two decimal places) in your Excel spreadsheet for the mean, standard deviation, standard error of the mean, and for both the lower limit and the upper limit of your confidence interval. If you printed this spreadsheet now, the lower limit of the confidence interval (26.92) and the upper limit of the confidence interval (29.42) would "dribble over" onto a second printed page because the information on the spreadsheet is too large to fit onto one page in its present format.

So, you need to use Excel's "Scale to Fit" commands that we discussed in Chap. 2 (see Sect. 2.4) to reduce the size of the spreadsheet to 95 % of its current size using the Page Layout/Scale to Fit function. Do that now, and notice that the dotted line to the right of 26.92 and 29.42 indicates that these numbers would now fit onto one page when the spreadsheet is printed out (see Fig. 3.4)

Chevy Impala					
Miles per gallon					
30.9					
24.5		n	25		
31.2					
28.7					
35.1		Mean	28.17		
29.0					
28.8					
23.1		STDEV	3.03		
31.0					
30.2					
28.4		s.e	0.61		
29.3					
24.2					
27.0		95% confidence interval			
26.7					
31.0			Lower limit:		26.92
23.5					
29.4			Upper Limit:		29.42
26.3					
27.5					
28.2	Draw a picture below this confidence interval				
28.4					
29.1	26.92	--------- 28 --------- 28.17 -------------		29.42	
21.9	lower	ref. Mean		upper	
30.9	limit	value		limit	
	Conclusion:				

Fig. 3.4 Result of Using the TINV Function to Find the Confidence Interval About the Mean

Note that you have drawn a picture of the 95 % confidence interval beneath cell B26, including the lower limit, the upper limit, the mean, and the reference value of 28 mpg given in the claim that the company wants to make about the car's miles per gallon performance.

Now, let's write the conclusion to your research study on your spreadsheet:

C33: Since the reference value of 28 is inside
C34: the confidence interval, we accept that
C35: the Chevy Impala does get 28 mpg.

Your research study accepted the claim that the Chevy Impala did get 28 miles per gallon. The average miles per gallon in your study was 28.17 (see Fig. 3.5).

Chevy Impala					
Miles per gallon					
30.9					
24.5		n	25		
31.2					
28.7					
35.1		Mean	28.17		
29.0					
28.8					
23.1		STDEV	3.03		
31.0					
30.2					
28.4		s.e	0.61		
29.3					
24.2					
27.0		95% confidence interval			
26.7					
31.0			Lower limit:		26.92
23.5					
29.4			Upper Limit:		29.42
26.3					
27.5					
28.2	Draw a picture below this confidence interval				
28.4					
29.1	26.92	--------- 28 --------- 28.17 --------------		29.42	
21.9	lower	ref. Mean		upper	
30.9	limit	value		limit	
	Conclusion:	Since the reference value of 28 is inside			
		the confidence interval, we accept that			
		the Chevy Impala does get 28 mpg.			

Fig. 3.5 Final Spreadsheet for the Chevy Impala Confidence Interval About the Mean

Save your resulting spreadsheet as: CHEVY7

3.2 Hypothesis Testing

One of the important activities of social science researchers is that they attempt to "check" their assumptions about the world by testing these assumptions in the form of hypotheses.

A typical hypothesis is in the form: "*If x, then y.*"

Some examples would be:

1. "If we use this new method of teaching reading in 4th grade classes, the Iowa Test of Basic Skills (ITBS) grade equivalent (GE) reading scores will improve by 5 %."

2. "If we raise our price by 5 %, then our sales dollars for this product will decrease by 4 % but our profit for this product will increase by 3 %."
3. "If we use this new method of teaching mathematics to ninth graders in algebra, then our math achievement scores will go up by 10 %."
4. "If we change the format for teaching Introductory Psychology to our undergraduates, then their final exam scores will increase by 10 %."

A hypothesis, then, to a social science researcher is a "guess" about what we think is true in the real world. We can test these guesses using statistical formulas to see if our predictions come true in the real world.

So, in order to perform these statistical tests, we must first state our hypotheses so that we can test our results against our hypotheses to see if our hypotheses match reality.

So, how do we generate hypotheses in social science research?

3.2.1 Hypotheses Always Refer to the Population of People or Events That You Are Studying

The first step is to understand that our hypotheses always refer to the *population* of people under study.

For example, if we are interested in studying 18–24 year-olds in St. Louis as our target market, and we select a sample of people in this age group in St. Louis, depending on how we select our sample, we are hoping that our results of this study are useful in generalizing our findings to *all* 18–24 year-olds in St. Louis, and not just to the particular people in our sample.

The entire group of 18–24 year-olds in St. Louis would be the *population* that we are interested in studying, while the particular group of people in our study are called the *sample* from this population.

Since our sample sizes typically contain only a few people, we interested in the results of our sample *only insofar as the results of our sample can be "generalized" to the population in which we are really interested.*

That is why our hypotheses always refer to the population, and never to the sample of people in our study.

You will recall from Chap. 1 that we used the symbol: $\overline{X}$ to refer to the mean of the sample we use in our research study (see Sect. 1.1).

We will use the symbol: μ (the Greek letter "mu") to refer to the *population mean.*

In testing our hypotheses, we are trying to decide which one of two competing hypotheses *about the population mean* we should accept given our data set.

3.2.2 *The Null Hypothesis and the Research (Alternative) Hypothesis*

These two hypotheses are called the *null hypothesis* and the *research hypothesis.*

Statistics textbooks typically refer to the *null hypothesis* with the notation: H_0.

The *research hypothesis* is typically referred to with the notation: H_1, and it is sometimes called the *alternative hypothesis.*

Let's explain first what is meant by the null hypothesis and the research hypothesis:

(1) *The null hypothesis is what we accept as true unless we have compelling evidence that it is not true.*

(2) *The research hypothesis is what we accept as true whenever we reject the null hypothesis as true.*

This is similar to our legal system in America where we assume that a supposed criminal is innocent until he or she is proven guilty in the eyes of a jury. Our null hypothesis is that this defendant is innocent, while the research hypothesis is that he or she is guilty.

In the great state of Missouri, every license plate has the state slogan: "Show me." This means that people in Missouri think of themselves as not gullible enough to accept everything that someone says as true unless that person's actions indicate the truth of his or her claim. In other words, people in Missouri believe strongly that a person's actions speak much louder than that person's words.

Since both the null hypothesis and the research hypothesis cannot both be true, the task of hypothesis testing using statistical formulas is to decide which one you will accept as true, and which one you will reject as true.

Sometimes in social science research a series of rating scales is used to measure people's attitudes toward a company, toward one of its products, or toward their intention-to-buy that company's products. These rating scales are typically 5-point, 7-point, or 10-point scales, although other scale values are often used as well.

3.2.2.1 Determining the Null Hypothesis and the Research Hypothesis When Rating Scales Are Used

Here is a typical example of a 7-point scale in education for parents of 8th grade pupils at the end of a school year (see Fig. 3.6):

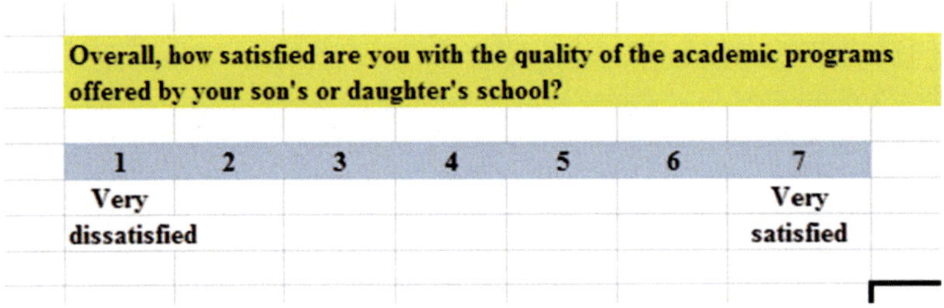

Fig. 3.6 Example of a Rating Scale Item for Parents of 8th Graders (Practical Example)

So, how do we decide what to use as the null hypothesis and the research hypothesis whenever rating scales are used?

Objective: To decide on the null hypothesis and the research hypothesis whenever rating scales are used.

In order to make this determination, we will use a simple rule:

Rule: Whenever rating scales are used, we will use the "middle" of the scale as the null hypothesis and the research hypothesis

In the above example, since 4 is the number in the middle of the scale (i.e., three numbers are below it, and three numbers are above it), our hypotheses become:

Null hypothesis : $\mu = 4$
Research hypothesis : $\mu \neq 4$

In the above rating scale example, if the result of our statistical test for this one attitude scale item indicates that our population mean is "close to 4," we say that we accept the null hypothesis that the parents of 8th grade pupils were neither satisfied nor dissatisfied with the quality of the academic programs offered by their son's or daughter's school.

In the above example, if the result of our statistical test indicates that the population mean is significantly different from 4, we reject the null hypothesis and accept the research hypothesis *by stating either that*:

"*Parents of* 8th *grade pupils were significantly satisfied with the quality of the academic programs offered by their son's or daughter's school*" (this is true whenever our sample mean is significantly greater than our expected population mean of 4).

or

"*Parents of* 8th *grade pupils were significantly dissatisfied with the quality of the academic programs offered by their son's or daughter's school*" (this is accepted as true whenever our sample mean is significantly less than our expected population mean of 4).

Both of these conclusions cannot be true. We accept one of the hypotheses as "true" based on the data set in our research study, and the other one as "not true" based on our data set.

The job of the social science researcher, then, is to decide which of these two hypotheses, the null hypothesis or the research hypothesis, he or she will accept as true given the data set in the research study.

Let's try some examples of rating scales so that you can practice figuring out what the null hypothesis and the research hypothesis are for each rating scale.

In the spaces in Fig. 3.7, write in the null hypothesis and the research hypothesis for the rating scales:

1. Webster University is an excellent university.

1	2	3	4	5
Strongly Disagree	Disagree	Undecided	Agree	Strongly Agree

Null hypothesis: μ = ______

Research hypothesis: μ ≠ ______

2. How would you rate the quality of teaching at Webster University?

poor	1	2	3	4	5	6	7	excellent

Null hypothesis: μ = ______

Research hypothesis: μ ≠ ______

3. How would you rate the quality of the faculty at Webster University?

1	2	3	4	5	6	7	8	9	10
very poor									very good

Null hypothesis: μ = ______

Research hypothesis: μ ≠ ______

Fig. 3.7 Examples of Rating Scales for Determining the Null Hypothesis and the Research Hypothesis

How did you do?

Here are the answers to these three questions:

1. The null hypothesis is $\mu = 3$, and the research hypothesis is $\mu \neq 3$ on this 5-point scale (i.e. the "middle" of the scale is 3).
2. The null hypothesis is $\mu = 4$, and the research hypothesis is $\mu \neq 4$ on this 7-point scale (i.e., the "middle" of the scale is 4).
3. The null hypothesis is $\mu = 5.5$, and the research hypothesis is $\mu \neq 5.5$ on this 10-point scale (i.e., the "middle" of the scale is 5.5 since there are 5 numbers below 5.5 and 5 numbers above 5.5).

As another example, Holiday Inn Express in its Stay Smart Experience Survey uses 4-point scales where:

1 = Not So Good
2 = Average
3 = Very Good
4 = Great

On this scale, the null hypothesis is: $\mu = 2.5$ and the research hypothesis is: $\mu \neq 2.5$, because there are two numbers below 2.5, and two numbers above 2.5 on that rating scale.

Now, let's discuss the 7 STEPS of hypothesis testing for using the confidence interval about the mean.

3.2.3 *The 7 Steps for Hypothesis-Testing Using the Confidence Interval About the Mean*

Objective: To learn the 7 steps of hypothesis-testing using the confidence interval about the mean

There are seven basic steps of hypothesis-testing for this statistical test.

3.2.3.1 STEP 1: State the Null Hypothesis and the Research Hypothesis

If you are using numerical scales in your survey, you need to remember that these hypotheses refer to the "middle" of the numerical scale. For example, if you are using 7-point scales with 1 = poor and 7 = excellent, these hypotheses would refer to the middle of these scales and would be:

Null hypothesis H_0: $\mu = 4$
Research hypothesis H_1: $\mu \neq 4$

3.2.3.2 STEP 2: Select the Appropriate Statistical Test

In this chapter we are studying the confidence interval about the mean, and so we will select that test.

3.2.3.3 STEP 3: Calculate the Formula for the Statistical Test

You will recall (see Sect. 3.1.5) that the formula for the confidence interval about the mean is:

$$\overline{X} \pm \text{t s.e.} \tag{3.2}$$

We discussed the procedure for computing this formula for the confidence interval about the mean using Excel earlier in this chapter, and the steps involved in using that formula are:

1. Use Excel's =COUNT function to find the sample size.
2. Use Excel's =AVERAGE function to find the sample mean, $\overline{X}$.
3. Use Excel's =STDEVfunction to find the standard deviation, STDEV.
4. Find the standard error of the mean (s.e.) by dividing the standard deviation (STDEV) by the square root of the sample size, n.
5. Use Excel's TINV function to find the lower limit of the confidence interval.
6. Use Excel's TINV function to find the upper limit of the confidence interval.

3.2.3.4 STEP 4: Draw a Picture of the Confidence Interval About the Mean, Including the Mean, the Lower Limit of the Interval, the Upper Limit of the Interval, and the Reference Value Given in the Null Hypothesis, H_0

3.2.3.5 STEP 5: Decide on a Decision Rule

(a) *If the reference value is inside the confidence interval, accept the null hypothesis, H_0*
(b) *If the reference value is outside the confidence interval, reject the null hypothesis, H_0, and accept the research hypothesis, H_1*

3.2.3.6 STEP 6: State the Result of Your Statistical Test

There are two possible results when you use the confidence interval about the mean, and only one of them can be accepted as "true." So your result would be one of the following:

Either: Since the reference value is inside the confidence interval, *we accept the null hypothesis, H_0*

Or: Since the reference value is outside the confidence interval, *we reject the null hypothesis, H_0, and accept the research hypothesis, H_1*

3.2.3.7 STEP 7: State the Conclusion of Your Statistical Test in Plain English!

In practice, this is more difficult than it sounds because you are trying to summarize the result of your statistical test in simple English that is both concise and accurate so that someone who has never had a statistics course (such as your boss, perhaps) can understand the conclusion of your test. This is a difficult task, and we will give you lots of practice doing this last and most important step throughout this book.

Objective: To write the conclusion of the confidence interval about the mean test

Let's set some basic rules for sating the conclusion of a hypothesis test.

Rule #1: *Whenever you reject H_0 and accept H_1, you must use the word "significantly" in the conclusion to alert the reader that this test found an important result.*

Rule #2: *Create an outline in words of the "key terms" you want to include in your conclusion so that you do not forget to include some of them.*

Rule #3: *Write the conclusion in plain English so that the reader can understand it even if that reader has never taken a statistics course.*

Let's practice these rules using the Chevy Impala Excel spreadsheet that you created earlier in this chapter, but first we need to state the hypotheses for that car.

If General Motors wants to claim that the Chevy Impala gets 28 miles per gallon on a billboard ad, the hypotheses would be:

$$H_0: \quad \mu = 28\,\text{mpg}$$
$$H_1: \quad \mu \neq 28\,\text{mpg}$$

You will remember that the reference value of 28 mpg was inside the 95 % confidence interval about the mean for your data, so we would accept H_0 for the Chevy Impala that the car does get 28 mpg.

Objective: To state the result when you accept H_0

Result: Since the reference value of 28 mpg is inside the confidence interval, we accept the null hypothesis, H_0

Let's try our three rules now:

Objective: To write the conclusion when you accept H_0

Rule #1: *Since the reference value was inside the confidence interval, we cannot use the word "significantly" in the conclusion. This is a basic rule we are using in this chapter for every problem.*

Rule #2: The key terms in the conclusion would be:

- Chevy Impala
- reference value of 28 mpg

Rule #3: The Chevy Impala did get 28 mpg.

The process of writing the conclusion when you accept H_0 is relatively straightforward since you put into words what you said when you wrote the null hypothesis.

However, the process of stating the conclusion when you reject H_0 and accept H_1 is more difficult, so let's practice writing that type of conclusion with three practice case examples:

Objective: To write the result and conclusion when you reject H_0

CASE #1: Suppose that an ad in *The Wall Street Journal* claimed that the Honda Accord Sedan gets 34 miles per gallon. The hypotheses would be:

$H_0: \ \mu = 34\,\text{mpg}$
$H_1: \ \mu \neq 34\,\text{mpg}$

Suppose that your research yields the following confidence interval:

30	31	32	34
lower limit	Mean	upper limit	Ref. Value

Result: Since the reference value is outside the confidence interval, we reject the null hypothesis and accept the research hypothesis

The three rules for stating the conclusion would be:

Rule #1: We must include the word "significantly" since the reference value of 34 is outside the confidence interval.

Rule #2: The key terms would be:

- Honda Accord Sedan
- significantly
- either "more than" or "less than"
- and probably closer to

Rule #3: The Honda Accord Sedan got significantly less than 34 mpg, and it was probably closer to 31 mpg.

Note that this conclusion says that the mpg was less than 34 mpg because the sample mean was only 31 mpg. Note, also, that when you find a significant result by rejecting the null hypothesis, *it is not sufficient to say only*: "*significantly less than 34 mpg*," because that does not tell the reader "how much less than 34 mpg" the sample mean was from 34 mpg. To make the conclusion clear, you need to add: "probably closer to 31 mpg" since the sample mean was only 31 mpg.

CASE #2: Suppose that you have been hired as a consultant by the St. Louis Symphony Orchestra (SLSO) to analyze the data from an Internet survey of attendees for a concert in Powell Symphony Hall in St. Louis last month. You have decided to practice your data analysis skills on Question #7 given in Fig. 3.8:

Question #7: **"Overall, how satisfied have you been with your experience(s) at SLSO concerts?"**

1	2	3	4	5	6	7
Extremely dissatisfied						Extremely satisfied

Fig. 3.8 Example of a Survey Item Used by the St. Louis Symphony Orchestra (SLSO)

The hypotheses for this one item would be:

$H_0: \quad \mu = 4$
$H_1: \quad \mu \neq 4$

Essentially, the null hypothesis equal to 4 states that if the obtained mean score for this question is not significantly different from 4 on the rating scale, then attendees, overall, were neither satisfied nor dissatisfied with their SLSO concerts.

Suppose that your analysis produced the following confidence interval for this item on the survey.

1.8	2.8	3.8	4
lower limit	Mean	upper limit	Ref. Value

Result: *Since the reference value is outside the confidence interval, we reject the null hypothesis and accept the research hypothesis.*

Rule #1: You must include the word "significantly" since the reference value is outside the confidence interval

Rule #2: The key terms would be:

- attendees
- SLSO Internet survey
- significantly
- last month
- either satisfied or dissatisfied (since the result is significant)
- experiences at concerts

Rule #3: Attendees were significantly dissatisfied, overall, on last month's Internet survey with their experiences at concerts of the SLSO.

Note that you need to use the word "dissatisfied" since the sample mean of 2.8 was on the dissatisfied side of the middle of the rating scale.

CASE #3: Suppose that a U.S. Senator from Missouri has asked his staff to conduct a phone survey of registered voters in his state to find out how they feel about his performance as a U.S. Senator. The Senator's staff then conducts a phone survey of a random sample of registered voters in Missouri using the survey item given in Fig. 3.9:

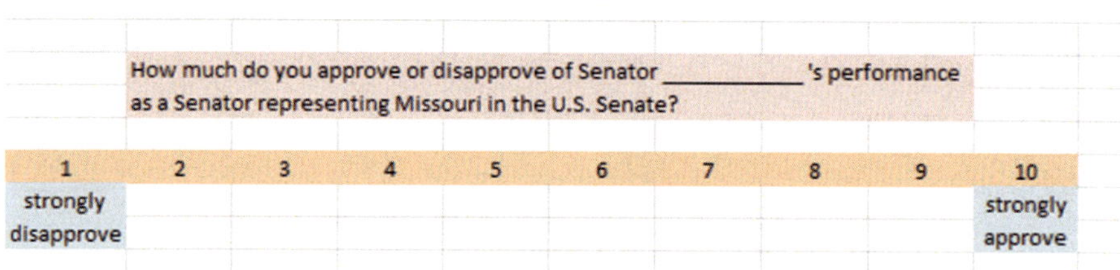

Fig. 3.9 Example of a Survey Item for a U.S. Senator from Missouri

This item would have the following hypotheses:

$H_0: \ \mu = 5.5$
$H_1: \ \mu \neq 5.5$

Suppose that your research produced the following confidence interval for this item on the survey:

5.5	5.7	5.8	5.9
Ref. Value	lower limit	Mean	upper limit

Result: Since the reference value is outside the confidence interval, we reject the null hypothesis and accept the research hypothesis

The three rules for stating the conclusion would be:

Rule #1: You must include the word "significantly" since the reference value is outside the confidence interval

Rule #2: The key terms would be:

- registered voters in Missouri
- significantly
- either "approve" or "disapprove"
- Senator ______'s
- performance in the U.S. Senate
- phone survey

Rule #3: Registered voters in Missouri significantly approved of Senator ______'s performance in the U.S. Senate in a phone survey.

Since the mean rating of the Senator's performance (5.8) was significantly greater than 5.5 on the approve side of the scale, we would say "significantly approved" to indicate this fact.

The three practice problems at the end of this chapter will give you additional practice in stating the conclusion of your result, and this book will include many more examples that will help you to write a clear and accurate conclusion to your research findings.

3.3 Alternative Ways to Summarize the Result of a Hypothesis Test

It is important for you to understand that in this book we are summarizing an hypothesis test in one of two ways: (1) We accept the null hypothesis, or (2) We reject the null hypothesis and accept the research hypothesis. We are consistent in the use of these words so that you can understand the concept underlying hypothesis testing.

However, there are many other ways to summarize the result of an hypothesis test, and all of them are correct theoretically, even though the terminology differs. If you are taking a course with a professor who wants you to summarize the results of a statistical test of hypotheses in language which is different from the language we are using in this book, do not panic! If you understand the concept of hypothesis testing as described in this book, you can then translate your understanding to use the terms that your professor wants you to use to reach the same conclusion to the hypothesis test.

Statisticians and professors of social science statistics all have their own language that they like to use to summarize the results of an hypothesis test. There is no one set of words that these statisticians and professors will ever agree on, and so we have chosen the one that we believe to be easier to understand in terms of the concept of hypothesis testing.

To convince you that there are many ways to summarize the results of an hypothesis test, we present the following quotes from prominent statistics and research books to give you an idea of the different ways that are possible.

3.3.1 Different Ways to Accept the Null Hypothesis

The following quotes are typical of the language used in statistics and research books when the null hypothesis is accepted:

"The null hypothesis is not rejected." (Black 2010, p. 310)
"The null hypothesis cannot be rejected." (McDaniel and Gates 2010, p. 545)
"The null hypothesis ... claims that there is no difference between groups." (Salkind 2010, p. 193)
"The difference is not statistically significant." (McDaniel and Gates 2010, p. 545)
"... the obtained value is not extreme enough for us to say that the difference between Groups 1 and 2 occurred by anything other than chance." (Salkind 2010, p. 225)
"If we do not reject the null hypothesis, we conclude that there is not enough statistical evidence to infer that the alternative (hypothesis) is true." (Keller 2009, p. 358)
"The research hypothesis is not supported." (Zikmund and Babin 2010, p. 552)

3.3.2 Different Ways to Reject the Null Hypothesis

The following quotes are typical of the quotes used in statistics and research books when the null hypothesis is rejected:

"The null hypothesis is rejected." (McDaniel and Gates 2010, p. 546)
"If we reject the null hypothesis, we conclude that there is enough statistical evidence to infer that the alternative hypothesis is true." (Keller 2009, p. 358)
"If the test statistic's value is inconsistent with the null hypothesis, we reject the null hypothesis and infer that the alternative hypothesis is true." (Keller 2009, p. 348)
"Because the observed value ... is greater than the critical value ..., the decision is to reject the null hypothesis." (Black 2010, p. 359)
"If the obtained value is more extreme than the critical value, the null hypothesis cannot be accepted." (Salkind 2010, p. 243)
"The critical t-value ... must be surpassed by the observed t-value if the hypothesis test is to be statistically significant ...." (Zikmund and Babin 2010, p. 567)

"The calculated test statistic . . . exceeds the upper boundary and falls into this rejection region. The null hypothesis is rejected." (Weiers 2011, p. 330)

You should note that all of the above quotes are used by statisticians and professors when discussing the results of an hypothesis test, and so you should not be surprised if someone asks you to summarize the results of a statistical test using a different language than the one we are using in this book.

3.4 End-of-Chapter Practice Problems

1. Health Care Reform is a "hot topic" in the U.S. with strong opinions as to how much the federal government should be involved in this issue. Some people favor a private medical insurance plan while still others favor a federal government-funded medical insurance plan. Suppose that you have been asked to write a survey on this topic that would be used in a mail survey of registered

HEALTH CARE REFORM SURVEY

Item #10: "What type of health care program should be adopted in the U.S.?"

0	1	2	3	4	5	6	7	8	9	10
favor a private medical insurance plan					undecided			favor a government-funded medical insurance plan		

Data
3
6
4
5
7
9
1
4
2
6
4
7
9
10
5
3
8

Fig. 3.10 Worksheet Data for Chap. 3: Practice Problem #1

voters in the state of Illinois. Suppose, further, that after developing this survey, you ran a pilot test with a small sample of registered voters to test your Excel skills. The hypothetical datafor Item #10 of this survey appear in Fig. 3.10.

Important note: Be careful! Is this a 10-point scale, or an 11-point scale? It makes a big difference in how you state your hypotheses!

(a) To the right of this table, use Excel to find the sample size, mean, standard deviation, and standard error of the mean for the figures. Label your answers. Use number format (two decimal places) for the mean, standard deviation, and standard error of the mean.
(b) Enter the null hypothesis and the research hypothesis onto your spreadsheet.
(c) Use Excel's TINV function to find the 95 % confidence interval about the mean for these figures. Label your answers. Use number format (two decimal places).
(d) Enter your *result* onto your spreadsheet.
(e) Enter your *conclusion in plain English* onto your spreadsheet.
(f) Print the final spreadsheet to fit onto one page (if you need help remembering how to do this, see the objectives at the end of Chap. 2 in Sect. 2.4)
(g) On your printout, draw a diagram of this 95 % confidence interval by hand
(h) Save the file as: Health3

2. Different special interest groups have strong opinions about what percentage of the U.S. federal budget should be devoted to their interests. Suppose that a U.S. congress woman in your voting district asked you to create a survey that could be mailed to registered voters in her voting district to obtain their opinions about the federal budget for various programs. After creating the survey, you decide to run a pilot test on a small sample of registered voters to test your Excel

SPENDING ON DEFENSE IN THE U.S. NATIONAL BUDGET

Item #7: "How would you rate U.S. government spending on the defense budget as a percent of the U.S. national budget?

1	2	3	4	5	6	7
should be decreased			should remain about the same			should be increased

Data
3
4
5
3
6
2
5
1
4
2
4
3
3
2
4
3
3
4
1

Fig. 3.11 Worksheet Data for Chap. 3: Practice Problem #2

skills. You select a random sample of voters and the hypothetical data from Item #7 are given in Fig. 3.11.

Create an Excel spreadsheet with these data.

(a) Use Excel to the right of the table to find the sample size, mean, standard deviation, and standard error of the mean for these data. Label your answers, and use two decimal places for the mean, standard deviation, and standard error of the mean

(b) Enter the null hypothesis and the research hypothesis for this item on your spreadsheet.

(c) Use Excel's TINV function to find the 95 % confidence interval about the mean for these data. Label your answers on your spreadsheet. Use two decimal places for the lower limit and the upper limit of the confidence interval.

(d) Enter the *result* of the test on your spreadsheet.
(e) Enter the *conclusion* of the test in plain English on your spreadsheet.
(f) Print your final spreadsheet so that it fits onto one page (if you need help remembering how to do this, see the objectives at the end of Chap. 2 in Sect. 2.4).
(g) Draw a picture of the confidence interval, including the reference value, onto your spreadsheet.
(h) Save the final spreadsheet as: Defense14

3. Welch's sells a small can of what the company claims as "100 % Grape Juice" and the package states that the can contains 5.5 fluid ounces (FL.OZ.) of grape juice, and the can also labels this amount as 163 ml of grape juice. Suppose that you have been asked to take a random sample of cans produced today to see if the cans contained 163 ml of grape juice. You select a random sample of cans, and the hypothetical results are given in Fig. 3.12:

WELCH'S 100% GRAPE JUICE

Research question: "Does the average can of Welch's 100% Grape Juice produced today contain 163 ml of grape juice?"

ml
165
158
163
159
154
157
159
161
164
154
157
161
163

Fig. 3.12 Worksheet Data for Chap. 3: Practice Problem #3

Create an Excel spreadsheet with these data.

(a) Use Excel to the right of the table to find the sample size, mean, standard deviation, and standard error of the mean for these data. Label your answers, and use two decimal places and number format for the mean, standard deviation, and standard error of the mean

(b) Enter the null hypothesis and the research hypothesis for this item onto your spreadsheet.
(c) Use Excel's TINV function to find the 95 % confidence interval about the mean for these data. Label your answers on your spreadsheet. Use two decimal places in number format for the lower limit and the upper limit of the confidence interval.
(d) Enter the *result* of the test on your spreadsheet.
(e) Enter the *conclusion* of the test in plain English on your spreadsheet.
(f) Print your final spreadsheet so that it fits onto one page (if you need help remembering how to do this, see the objectives at the end of Chap. 2 in Sect. 2.4).
(g) Draw a picture of the confidence interval, including the reference value, onto your spreadsheet.
(h) Save the final spreadsheet as: grape3

References

Black, K. Business Statistics: for Contemporary Decision Making (6th ed.). Hoboken, NJ: John Wiley& Sons, Inc., 2010.

Frankfort-Nachmias, C. and Nachmias, D. Research Methods in the Social Sciences (7th ed.). New York, NY: Worth Publishers, 2008.

Keller, G. Statistics for Management and Economics (8th ed.). Mason, OH: South-Western Cengage Learning, 2009.

King, G., Keohane, R.O., and Verba, S. Designing Social Inquiry: Scientific Inference in Qualitative Research. Princeton, NJ: Princeton University Press, 1994.

Levine, D.M. Statistics for Managers using Microsoft Excel (6th ed.). Boston, MA: Prentice Hall/ Pearson, 2011.

McDaniel, C. and Gates, R. Marketing Research (8th ed.). Hoboken, NJ: John Wiley & Sons, Inc., 2010.

Pollock, P.H. III. The Essentials of Political Analysis (3rd ed.). Washington, D.C.: CQ Press, 2009.

Salkind, N.J. Statistics for People Who (think they) Hate Statistics (2nd Excel 2007 ed.). Los Angeles, CA: Sage Publications, 2010.

Steinberg, W.J. Statistics Alive! Thousand Oaks, CA: Sage Publications, 2008.

Weiers, R.M. Introduction to Business Statistics (7th ed.). Mason, OH: South-Western Cengage Learning, 2011.

Zikmund, W.G. and Babin, B.J. Exploring Marketing Research (10th ed.). Mason, OH: South-Western Cengage Learning, 2010.

Chapter 4
One-Group t-Test for the Mean

In this chapter, you will learn how to use one of the most popular and most helpful statistical tests in social science research: the one-group t-test for the mean.

The formula for the one-group t-test is as follows:

$$t = \frac{\overline{X} - \mu}{S_{\overline{X}}} \text{ where} \tag{4.1}$$

$$\text{s.e.} = S_{\overline{X}} = \frac{S}{\sqrt{n}} \tag{4.2}$$

This formula asks you to take the mean ($\overline{X}$) and subtract the population mean (μ) from it, and then divide the answer by the standard error of the mean (s.e.). The standard error of the mean equals the standard deviation divided by the square root of n (the sample size).

Let's discuss the 7 STEPS of hypothesis testing using the one-group t-test so that you can understand how this test is used.

4.1 The 7 STEPS for Hypothesis-Testing Using the One-Group t-Test

Objective: To learn the 7 steps of hypothesis-testing using the one-group t-test

Before you can try out your Excel skills on the one-group t-test, you need to learn the basic steps of hypothesis-testing for this statistical test. There are 7 steps in this process:

© Springer International Publishing Switzerland 2016

T.J. Quirk, *Excel 2016 for Social Science Statistics*, Excel for Statistics,
DOI 10.1007/978-3-319-39711-5_4

4.1.1 STEP 1: State the Null Hypothesis and the Research Hypothesis

If you are using numerical scales in your survey, you need to remember that these hypotheses refer to the "middle" of the numerical scale. For example, if you are using 7-point scales with 1 = poor and 7 = excellent, these hypotheses would refer to the middle of these scales and would be:

$$\text{Null hypothesis } H_0: \quad \mu = 4$$
$$\text{Research hypothesis } H_1: \quad \mu \neq 4$$

As a second example, suppose that you worked for Honda Motor Company and that you wanted to place a magazine ad that claimed that the new Honda Fit got 35 miles per gallon (mpg). The hypotheses for testing this claim on actual data would be:

$$H_0: \quad \mu = 35\,\text{mpg}$$
$$H_1: \quad \mu \neq 35\text{mpg}$$

4.1.2 STEP 2: Select the Appropriate Statistical Test

In this chapter we will be studying the one-group t-test, and so we will select that test.

4.1.3 STEP 3: Decide on a Decision Rule for the One-Group t-Test

(a) If the absolute value of t is less than the critical value of t, accept the null hypothesis.
(b) If the absolute value of t is greater than the critical value of t, reject the null hypothesis and accept the research hypothesis.

You are probably saying to yourself: "That sounds fine, but how do I find the absolute value of t?"

4.1.3.1 Finding the Absolute Value of a Number

To do that, we need another objective:

Objective: To find the absolute value of a number

If you took a basic algebra course in high school, you may remember the concept of "absolute value." In mathematical terms, the absolute value of any number is *always* that number expressed as a positive number.

For example, the absolute value of 2.35 is +2.35.

And the absolute value of minus 2.35 (i.e. −2.35) is also +2.35.

This becomes important when you are using the t-table in Appendix E of this book. We will discuss this table later when we get to Step 5 of the one-group t-test where we explain how to find the critical value of t using Appendix E.

4.1.4 STEP 4: Calculate the Formula for the One-Group t-Test

Objective: To learn how to use the formula for the one-group t-test

The formula for the one-group t-test is as follows:

$$t = \frac{\overline{X} - \mu}{S_{\overline{X}}} \text{ where} \tag{4.1}$$

$$\text{s.e.} = S_{\overline{X}} = \frac{S}{\sqrt{n}} \tag{4.2}$$

This formula makes the following assumptions about the data (Foster et al. 1998): (1) The data are independent of each other (i.e., each person receives only one score), (2) the *population* of the data is normally distributed, and (3) the data have a constant variance (note that the standard deviation is the square root of the variance).

To use this formula, you need to follow these steps:

1. Take the sample mean in your research study and subtract the population mean μ from it (remember that the population mean for a study involving numerical rating scales is the "middle" number in the scale).
2. Then take your answer from the above step, and divide your answer by the standard error of the mean for your research study (you will remember that you learned how to find the standard error of the mean in Chap. 1; to find the standard error of the mean, just take the standard deviation of your research study and divide it by the square root of n, where n is the number of people used in your research study).
3. The number you get after you complete the above step is the value for t that results when you use the formula stated above.

4.1.5 STEP 5: Find the Critical Value of t in the t-Table in Appendix E

Objective: To find the critical value of t in the t-table in Appendix E

Before we get into an explanation of what is meant by "the critical value of t," let's give you practice in finding the critical value of t by using the t-table in Appendix E.

Keep your finger on Appendix E as we explain how you need to "read" that table.

Since the test in this chapter is called the "one-group t-test," you will use the first column on the left in Appendix E to find the critical value of t for your research study (note that this column is headed: "sample size n").

To find the critical value of t, you go down this first column until you find the sample size in your research study, and then you go to the right and read the critical value of t for that sample size in the critical t column in the table (note that *this is the column that you would use for both the one-group t-test and the 95 % confidence interval about the mean*).

For example, if you have 27 people in your research study, the critical value of t is 2.056.

If you have 38 people in your research study, the critical value of t is 2.026.

If you have more than 40 people in your research study, the critical value of t is always 1.96.

Note that the "critical t column" in Appendix E represents the value of t that you need to obtain to be 95 % confident of your results as "significant" results.

The critical value of t is the value that tells you whether or not you have found a "significant result" in your statistical test.

The t-table in Appendix E represents a series of "bell-shaped normal curves" (they are called bell-shaped because they look like the outline of the Liberty Bell that you can see in Philadelphia outside of Independence Hall).

The "middle" of these normal curves is treated as if it were zero point on the x-axis (the technical explanation of this fact is beyond the scope of this book, but any good statistics book (e.g. Zikmund and Babin 2010) will explain this concept to you if you are interested in learning more about it).

Thus, values of t that are to the right of this zero point are positive values that use a plus sign before them, and values of t that are to the left of this zero point are negative values that use a minus sign before them. Thus, some values of t are positive, and some are negative.

However, every statistics book that includes a t-table only reprints the *positive* side of the t-curves because the negative side is the mirror image of the positive side; this means that the negative side contains the exact same numbers as the positive side, but the negative numbers all have a minus sign in front of them.

Therefore, to use the t-table in Appendix E, you need to *take the absolute value of the t-value you found when you use the t-test formula* since the t-table in Appendix E only has the values of t that are the positive values for t.

Throughout this book, we are assuming that you want to be 95 % confident in the results of your statistical tests. Therefore, the value for t in the t-table in Appendix E tells you whether or not the t-value you obtained when you used the formula for the one-group t-test is within the 95 % interval of the t-curve range that that t-value would be expected to occur with 95 % confidence.

If the t-value you obtained when you used the formula for the one-group t-test is *inside* of the 95 % confidence range, we say that the result you found is *not significant* (note that this is equivalent to *accepting the null hypothesis!).*

If the t-value you found when you used the formula for the one-group t-test is *outside* of this 95 % confidence range, we say that you have found a *significant result* that would be expected to occur less than 5 % of the time (note that this is equivalent to *rejecting the null hypothesis and accepting the research hypothesis*).

4.1.6 STEP 6: State the Result of Your Statistical Test

There are two possible results when you use the one-group t-test, and only one of them can be accepted as "true."

Either: Since the absolute value of t that you found in the t-test formula is *less than the critical value of t* in Appendix E, you accept the null hypothesis.

Or: Since the absolute value of t that you found in the t-test formula is *greater than the critical value of t* in Appendix E, you reject the null hypothesis, and accept the research hypothesis.

4.1.7 STEP 7: State the Conclusion of Your Statistical Test in Plain English!

In practice, this is more difficult than it sounds because you are trying to summarize the result of your statistical test in simple English that is both concise and accurate so that someone who has never had a statistics course (such as your boss, perhaps) can understand the result of your test. This is a difficult task, and we will give you lots of practice doing this last and most important step throughout this book.

If you have read this far, you are ready to sit down at your computer and perform the one-group t-test using Excel on some hypothetical data from the Guest Satisfaction Survey used by Marriott Hotels.

Let's give this a try.

4.2 One-Group t-Test for the Mean

Suppose that you have been hired as a statistical consultant by Marriott Hotel in St. Louis to analyze the data from a Guest Satisfaction survey that they give to all customers to determine the degree of satisfaction of these customers for various activities of the hotel.

The survey contains a number of items, but suppose item #7 is the one in Fig. 4.1:

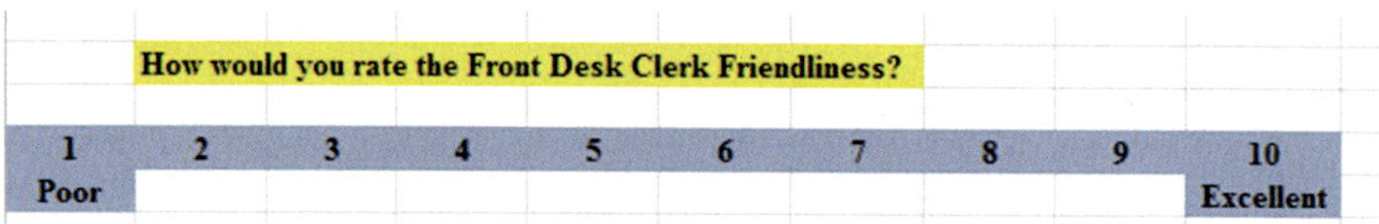

Fig. 4.1 Sample Survey Item for Marriot Hotel (Practical Example)

Suppose further, that you have decided to analyze the data from last week's customers using the one-group t-test.

Important note: You would need to use this test for each of the survey items separately.

Suppose that the hypothetical data for Item #7 from last week at the St. Louis Marriott Hotel were based on a sample size of 124 guests who had a mean score on this item of 6.58 and a standard deviation on this item of 2.44.

> Objective: To analyze the data for each question separately using the one-group t-test for each survey item.

Create an Excel spreadsheet with the following information:

B11: Null hypothesis:
B14: Research hypothesis:

Note: Remember that when you are using a rating scale item, both the null hypothesis and the research hypothesis refer to the "middle of the scale." In the 10-point scale in this example, the middle of the scale is 5.5 since five numbers are below 5.5 (i.e., 1–5) and five numbers are above 5.5 (i.e. 6–10). Therefore, the hypotheses for this rating scale item are:

$H_0: \mu = 5.5$
$H_1: \mu \neq 5.5$

B17: n
B20: mean
B23: STDEV

B26: s.e.
B29: critical t
B32: t-test
B36: Result:
B41: Conclusion:

Now, use Excel:

D17: enter the sample size
D20: enter the mean
D23: enter the STDEV (see Fig. 4.2)

Fig. 4.2 Basic Data Table for Front Desk Clerk Friendliness

Null hypothesis:	
Research hypothesis:	
n	**124**
mean	**6.58**
STDEV	**2.44**
s.e.	
critical t	
t-test	
Result:	
Conclusion:	

D26: compute the standard error using the formula in Chap. 1
D29: find the critical t value of t in the t-table in Appendix E

Now, enter the following formula in cell D32 to find the t-test result:

$=(\mathrm{D20}-5.5)/\mathrm{D26}$ (no spaces between symbols)

This formula takes the sample mean (D20) and subtracts the population hypothesized mean of 5.5 from the sample mean, and THEN divides the answer by the standard error of the mean (D26). Note that you need to enter D20–5.5 with an open-parenthesis *before* D20 and a closed-parenthesis *after* 5.5 so that the *answer of 1.08 is THEN divided by the standard error of 0.22* to get the t-test result of 4.93.

Now, use two decimal places for both the s.e. and the t-test result (see Fig. 4.3).

Fig. 4.3 t-test Formula Result for Front Desk Clerk Friendliness

Null hypothesis:	
Research hypothesis:	
n	**124**
mean	**6.58**
STDEV	**2.44**
s.e.	**0.22**
critical t	**1.96**
t-test	**4.93**
Result:	
Conclusion:	

Now, write the following sentence in D36–D39 to summarize the result of the t-test:

D36: Since the absolute value of t of 4.93 is
D37: greater than the critical t of 1.96, we
D38: reject the null hypothesis and accept
D39: the research hypothesis.

Lastly, write the following sentence in D41–D43 to summarize the conclusion of the result for Item #7 of the Marriott Guest Satisfaction Survey:

D41: St. Louis Marriott Hotel guests rated the
D42: Front Desk Clerks as significantly
D43: friendly last week.

Save your file as: MARRIOTT3

Important note: You are probably wondering why we entered both the result and the conclusion in separate cells instead of in just one cell. This is because if you enter them in one cell, you will be very disappointed when you print out your final spreadsheet, because one of two things will happen that you will not like: (1) if you print the spreadsheet to fit onto only one page, the result and the conclusion will force the entire spreadsheet to be printed in such small font size that you will be unable to read it, or (2) if you do not print the final spreadsheet to fit onto one page, both the result and the conclusion will "dribble over" onto a second page instead of fitting the entire spread-sheet onto one page. In either case, your spreadsheet will not have a "professional look."

Print the final spreadsheet so that it fits onto one page as given in Fig. 4.4. Enter the null hypothesis and the research hypothesis by hand on your spreadsheet

Null hypothesis:		**μ = 5.5**	
Research hypothesis:		**μ ≠ 5.5**	
n	**124**		
mean	**6.58**		
STDEV	**2.44**		
s.e.	**0.22**		
critical t	**1.96**		
t-test	**4.93**		
Result:		**Since the absolute value of t of 4.93 is greater than the critical t of 1.96, we reject the null hypothesis and accept the research hypothesis.**	
Conclusion:		**St. Louis Marriott Hotel guests rated the Front Desk Clerks as significantly friendly last week.**	

Fig. 4.4 Final Spreadsheet for Front Desk Clerk Friendliness

Important note: It is important for you to understand that "technically" the above conclusion in statistical terms should state:

"St. Louis Marriott Hotel Guests rated the Front Desk Clerks as friendly last week, and this result was probably not obtained by chance."

However, throughout this book, we are using the term "significantly" in writing the conclusion of statistical tests to alert the reader that the result of the statistical test was probably not a chance finding, but instead of writing all of those words each time, we use the word "significantly" as a shorthand to the longer explanation. This makes it much easier for the reader to understand the conclusion when it is written "in plain English," instead of technical, statistical language.

For a more detailed explanation of the one-group t-test, see Pollock (2009) and Johnson and Reynolds (2008).

4.3 Can You Use Either the 95 Percent Confidence Interval About the Mean OR the One-Group t-Test When Testing Hypotheses?

You are probably asking yourself:

"It sounds like you could use *either* the 95 % confidence interval about the mean *or* the one-groupt-test to analyze the results of the types of problems described so far in this book? Is this a correct statement?"

The answer is a resounding: *"Yes!"*

Both the confidence interval about the mean and the one-group t-test are used often in social science research on the types of problems described so far in this book. *Both of these tests produce the same result and the same conclusion from the data set!*

Both of these tests are explained in this book because some researchers prefer the confidence interval about the mean test, others prefer the one-group t-test, and still others prefer to use both tests on the same data to make their results and conclusions clearer to the reader of their research reports. Since we do not know which of these tests your researcher prefers, we have explained both of them so that you are competent in the use of both tests in the analysis of statistical data.

Now, let's try your Excel skills on the one-group t-test on these three problems at the end of this chapter.

4.4 End-of-Chapter Practice Problems

1. Suppose that you are working as a research assistant on a research project for a political science professor who is interested in studying the political ideologies of undergraduates at her large state university. The professor has developed a survey, and Item #12 attempts to measure the degree of liberalism-conservatism of college students. Suppose, further than you have decided to test your Excel

skills on a small sample of students and to analyze the hypothetical data from a random sample of students in Fig. 4.5 for Question #12:

LIBERAL-CONSERVATIVISM OF COLLEGE STUDENTS						
Item #12:	"How would you rate your political ideology?"					
1	2	3	4	5	6	7
extermely liberal			moderate			extermely conservative
		Data				
		2				
		6				
		3				
		4				
		7				
		5				
		3				
		5				
		4				
		2				
		1				
		3				
		4				
		2				
		5				
		6				
		3				
		5				
		2				
		7				

Fig. 4.5 Worksheet Data for Chap. 4: Practice Problem #1

(a) Write the null hypothesis and the research hypothesis on your spreadsheet
(b) Use Excel to find the sample size, mean, standard deviation, and standard error of the mean to the right of the data set. Use number format (two decimal places) for the mean, standard deviation, and standard error of the mean.
(c) Enter the critical t from the t-table in Appendix E onto your spreadsheet, and label it.
(d) Use Excel to compute the t-value for these data (use two decimal places) and label it on your spreadsheet

(e) Type the result on your spreadsheet, and then type the conclusion in plain English on your spreadsheet
(f) Save the file as: Political13

2. Cultural anthropologists who are interested in studying human societies are known as ethnographers (Fetterman 2010). Suppose that an anthropologist has been asked to evaluate the effectiveness of a job-training program for adults, and that Item #18 of the survey being used asks the participants to express their opinion on the support and guidance they received during the program in terms of their securing employment. You are being asked to analyze the data for item #18, and so you decide to try out your Excel skills on a small sample of participants. The hypothetical data for a random sample of students from Item #18 are presented in Fig. 4.6:

JOB TRAINING PROGRAM SURVEY

Item #18: "How would you rate the support and guidance you received in this program in securing employment?"

1	2	3	4	5
poor		satisfactory		excellent

Data
3
5
2
4
3
5
4
5
3
4
2
5
1
5
4
5
3
5
4

Fig. 4.6 Worksheet Data for Chap. 4: Practice Problem #2

(a) *On your Excel spreadsheet*, write the null hypothesis and the research hypothesis for these data.
(b) Use Excel to find the *sample size, mean, standard deviation, and standard error of the mean* for these data (two decimal places for the mean, standard deviation, and standard error of the mean).
(c) Use Excel to perform a *one-group t-test* on these data (two decimal places).
(d) On your printout, type the *critical value of t* given in your t-table in Appendix E.
(e) On your spreadsheet, type the *result* of the t-test.
(f) On your spreadsheet, type the *conclusion* of your study in plain English.
(g) save the file as: Job3

3. Suppose that you have been hired as a marketing consultant by the Missouri Botanical Garden and have been asked to re-design the Comment Card survey that they have been asking visitors to The Garden to fill out after their visit. The Garden has been using a 5-point rating scale with 1 = poor and 5 = excellent. Suppose, further, that you have convinced The Garden staff to change to a 9-point scale with 1 = poor and 9 = excellent so that the data will have a larger standard deviation. The hypothetical results for Question #10 of your revised survey appear in Fig. 4.7.

MISSOURI BOTANICAL GARDEN								
VISITOR SURVEY								
Item #10: "How would you rate the helpfulness of The Garden staff?"								
1	2	3	4	5	6	7	8	9
poor								excellent
		Results						
		8						
		6						
		5						
		7						
		9						
		5						
		6						
		4						
		8						
		7						
		6						
		8						
		6						
		7						
		9						
		7						
		6						
		3						
		8						
		7						
		6						

Fig. 4.7 Worksheet Data for Chap. 4: Practice problem #3

(a) Write the null hypothesis and the research hypothesis on your spreadsheet
(b) Use Excel to find the sample size, mean, standard deviation, and standard error of the mean to the right of the data set. Use number format (two decimal places) for the mean, standard deviation, and standard error of the mean.
(c) Enter the critical t from the t-table in Appendix E onto your spreadsheet, and label it.
(d) Use Excel to compute the t-value for these data (use two decimal places) and label it on your spreadsheet
(e) Type the result on your spreadsheet, and then type the conclusion in plain English on your spreadsheet
(f) Save the file as: Garden5

References

Fetterman, D.M. Ethnography: Step-by-Step (3rd ed.). Los Angeles, CA: Sage Publications, 2010.

Foster, D.P., Stine, R.A., Waterman, R.P. Basic Business Statistics: A Casebook. New York, NY: Springer-Verlag, 1998.

Johnson, J.B. and H.T. Reynolds. Political Science Research Methods (6th ed.). Washington, D.C.: CQ Press, 2008.

Pollock, P.H. III. The Essentials of Political Analysis. (3rd ed.). Washington, D.C.: CQ Press, 2009.

Zikmund, W.G. and Babin, B.J. Exploring Marketing Research (10th ed.) Mason, OH: South-Western Cengage Learning, 2010.

Chapter 5
Two-Group t-Test of the Difference of the Means for Independent Groups

Up until now in this book, you have been dealing with the situation in which you have had only one group of people in your research study and only one measurement "number" on each of these people. We will now change gears and deal with the situation in which you are measuring two groups of people instead of only one group of people.

Whenever you have two completely different groups of people (i.e., no one person is in both groups, but every person is measured on only one variable to produce one "number" for each person), we say that the two groups are "independent of one another." This chapter deals with just that situation and that is why it is called the two-group t-test for independent groups.

The two assumptions underlying the two-group t-test are the following (Zikmund and Babin 2010): (1) both groups are sampled from a normal population, and (2) the variances of the two populations are approximately equal. Note that the standard deviation is merely the square root of the variance. (There are different formulas to use when each person is measured twice to create two groups of data, and this situation is called "dependent," but those formulas are beyond the scope of this book.) This book only deals with two groups that are independent of one another so that no person is in both groups of data.

When you are testing for the difference between the means for two groups, it is important to remember that there are two different formulas that you need to use depending on the sample sizes of the two groups:

(1) Use Formula #1 in this chapter when both of the groups have more than 30 people in them, and
(2) Use Formula #2 in this chapter when either one group, or both groups, have sample sizes less than 30 people in them.

We will illustrate both of these situations in this chapter.

But, first, we need to understand the steps involved in hypothesis-testing when two groups of people are involved before we dive into the formulas for this test.

© Springer International Publishing Switzerland 2016

T.J. Quirk, *Excel 2016 for Social Science Statistics*, Excel for Statistics,
DOI 10.1007/978-3-319-39711-5_5

5.1 The 9 STEPS for Hypothesis-Testing Using the Two-Group t-Test

Objective: To learn the 9 steps of hypothesis-testing using two groups of people and the two-group t-test

You will see that these steps parallel the steps used in the previous chapter that dealt with the one-group t-test, but there are some important differences between the steps that you need to understand clearly before we dive into the formulas for the two-group t-test.

5.1.1 STEP 1: Name One Group, Group 1, and the Other Group, Group 2

The formulas used in this chapter will use the numbers 1 and 2 to distinguish between the two groups. If you define which group is Group 1 and which group is Group 2, you can use these numbers in your computations without having to write out the names of the groups.

For example, if you are testing teenage boys on their preference for the taste of Coke or Pepsi, you could call the groups: "Coke" and "Pepsi." but this would require your writing out the words "Coke" or "Pepsi" whenever you wanted to refer to one of these groups. If you call the Coke group, Group 1, and the Pepsi group, Group 2, this makes it much easier to refer to the groups because it saves you writing time.

As a second example, you could be comparing the test market results for Kansas City versus Indianapolis, but if you had to write out the names of those cities whenever you wanted to refer to them, it would take you more time than it would if, instead, you named one city, Group 1, and the other city, Group 2.

Note, also, that it is completely arbitrary which group you call Group 1, and which Group you call Group 2. You will achieve the same result and the same conclusion from the formulas however you decide to define these two groups.

5.1.2 STEP 2: Create a Table That Summarizes the Sample Size, Mean Score, and Standard Deviation of Each Group

This step makes it easier for you to make sure that you are using the correct numbers in the formulas for the two-group t-test. If you get the numbers "mixed-up," your entire formula work will be incorrect and you will botch the problem terribly.

For example, suppose that you tested teenage boys on their preference for the taste of Coke versus Pepsi in which the boys were randomly assigned to taste just one of these brands and then rate its taste on a 100-point scale from 0 = poor to 100 = excellent. After the research study was completed, suppose that the Coke group had 52 boys in it, their mean taste rating was 55 with a standard deviation of 7, while the Pepsi group had 57 boys in it and their average taste rating was 64 with a standard deviation of 13.

The formulas for analyzing these data to determine if there was a significant different in the taste rating for teenage boys for these two brands require you to use six numbers correctly in the formulas: the sample size, the mean, and the standard deviation of each of the two groups. All six of these numbers must be used correctly in the formulas if you are to analyze the data correctly.

If you create a table to summarize these data, a good example of the table, using both Step 1 and Step 2, would be the data presented in Fig. 5.1:

Fig. 5.1 Basic Table Format for the Two-group t-test

Group	n	Mean	STDEV
1 (name it)			
2 (name it)			

For example, if you decide to call Group 1 the Coke group and Group 2 the Pepsi group, the following table would place the six numbers from your research study into the proper calls of the table as in Fig. 5.2:

Fig. 5.2 Results of Entering the Data Needed for the Two-group t-test

Group	n	Mean	STDEV
1 (name it)	52	55	7
2 (name it)	57	64	13

You can now use the formulas for the two-group t-test with more confidence that the six numbers will be placed in the proper place in the formulas.

Note that you could just as easily call Group 1 the Pepsi group and Group 2 the Coke group; it makes no difference how you decide to name the two groups; this decision is up to you.

5.1.3 STEP 3: State the Null Hypothesis and the Research Hypothesis for the Two-Group t-Test

If you have completed Step 1 above, this step is very easy because the null hypothesis and the research hypothesis will always be stated in the same way for the two-group t-test. The null hypothesis states that the population means of the two groups are equal, while the research hypothesis states that the population means of the two groups are not equal. In notation format, this becomes:

$H_0: \mu_1 = \mu_2$
$H_1: \mu_1 \neq \mu_2$

You can now see that this notation is much simpler than having to write out the names of the two groups in all of your formulas.

5.1.4 STEP 4: Select the Appropriate Statistical Test

Since this chapter deals with the situation in which you have two groups of people but only one measurement on each person in each group, we will use the two-group t-test throughout this chapter.

5.1.5 STEP 5: Decide on a Decision Rule for the Two-Group t-Test

The decision rule is exactly what it was in the previous chapter (see Sect. 4.1.3) when we dealt with the one-group t-test.

(a) If the absolute value of t is less than the critical value of t, accept the null hypothesis.
(b) If the absolute value of t is greater than the critical value of t, reject the null hypothesis and accept the research hypothesis.

Since you learned how to find the absolute value of t in the previous chapter (see Sect. 4.1.3.1), you can use that knowledge in this chapter.

5.1.6 *STEP 6: Calculate the Formula for the Two-Group t-Test*

Since we are using two different formulas in this chapter for the two-group t-test depending on the sample size of the people in the two groups, we will explain how to use those formulas later in this chapter.

5.1.7 *STEP 7: Find the Critical Value of t in the t-Table in Appendix E*

In the previous chapter where we were dealing with the one-group t-test, you found the critical value of t in the t-table in Appendix E by finding the sample size for the one group of people in the first column of the table, and then reading the critical value of t across from it on the right in the "critical t column" in the table (see Sect. 4.1.5). This process was fairly simple once you have had some practice in doing this step.

However, for the two-group t-test, the procedure for finding the critical value of t is more complicated because you have two different groups of people in your study, and they often have different sample sizes in each group.

To use Appendix E correctly in this chapter, you need to learn how to find the "degrees of freedom" for your study. We will discuss that process now.

5.1.7.1 Finding the Degrees of Freedom (df) for the Two-Group t-Test

Objective: To find the degrees of freedom for the two-group t-test and to use it to find the critical value of t in the t-table in Appendix E

The mathematical explanation of the concept of the "degrees of freedom" is beyond the scope of this book, but you can find out more about this concept by reading any good statistics book (e.g. Keller 2009). For our purposes, you can easily understand how to find the degrees of freedom and to use it to find the critical value of t in Appendix E. The formula for the degrees of freedom (df) is:

$$\text{degrees of freedom} = \text{df} = n_1 + n_2 - 2 \quad (5.1)$$

In other words, you add the sample size for Group 1 to the sample size for Group 2 and then subtract 2 from this total to get the number of degrees of freedom to use in Appendix E.

Take a look at Appendix E.

Instead of using the first column as we did in the one-group t-test that is based on the sample size, n, of one group of people, we need to use the second-column of this table (df) to find the critical value of t for the two-group t-test.

For example, if you had 13 people in Group 1 and 17 people in Group 2, the degrees of freedom would be: $13 + 17 - 2 = 28$, and the critical value of t would be 2.048 *since you look down the second column which contains the degrees of freedom* until you come to the number 28, and then read 2.048 in the "critical t column" in the table to find the critical value of t when df $= 28$.

As a second example, if you had 52 people in Group 1 and 57 people in Group 2, the degrees of freedom would be: $52 + 57 - 2 = 107$ When you go down the second column in Appendix E for the degrees of freedom, you find that *once you go beyond the degrees of freedom equal to 39, the critical value of t is always 1.96*, and that is the value you would use for the critical t with this example.

5.1.8 STEP 8: State the Result of Your Statistical Test

The result follows the exact same result format that you found for the one-group t-test in the previous chapter (see Sect. 4.1.6):

Either: Since the absolute value of t that you found in the t-test formula is *less than the critical value of t* in Appendix E, you accept the null hypothesis.

Or: Since the absolute value of t that you found in the t-test formula is *greater than the critical value of t* in Appendix E, you reject the null hypothesis and accept the research hypothesis.

5.1.9 STEP 9: State the Conclusion of Your Statistical Test in Plain English!

Writing the conclusion for the two-group t-test is more difficult than writing the conclusion for the one-group t-test because you have to decide what the difference was between the two groups.

When you accept the null hypothesis, the conclusion is simple to write: "There is no difference between the two groups in the variable that was measured."

But when you reject the null hypothesis and accept the research hypothesis, you need to be careful about writing the conclusion so that it is both accurate and concise.

Let's give you some practice in writing the conclusion of a two-group t-test.

5.1.9.1 Writing the Conclusion of the Two-Group t-Test When You Accept the Null Hypothesis

Objective: To write the conclusion of the two-group t-test when you have accepted the null hypothesis.

Suppose that you have been hired as a statistical consultant by Marriott Hotel in St. Louis to analyze the data from a Guest Satisfaction Survey that they give to all customers to determine the degree of satisfaction of these customers for various activities of the hotel.

The survey contains a number of items, but suppose Item #7 is the one in Fig. 5.3:

	How would you rate the Front Desk Clerk Friendliness?								
1	2	3	4	5	6	7	8	9	10
Poor									Excellent

Fig. 5.3 Marriott Hotel Guest Satisfaction Survey Item #7

Suppose further, that you have decided to analyze the data from last week's customers comparing men and women using the two-group t-test.

Important note: You would need to use this test for each of the survey items separately.

Suppose that the hypothetical data for Item #7 from last week at the St. Louis Marriott Hotel were based on a sample size of 124 men who had a mean score on this item of 6.58 and a standard deviation on this item of 2.44. Suppose that you also had data from 86 women from last week who had a mean score of 6.45 with a standard deviation of 1.86, and that you created the table given in Fig. 5.4.

Fig. 5.4 Worksheet Data for Males vs. Females for the St. Louis Marriott Hotel for Accepting the Null Hypothesis

Group	n	Mean	STDEV
1 Males	124	6.58	2.44
2 Females	86	6.45	1.86

We will explain later in this chapter how to produce the results of the two-group t-test using its formulas, but, for now, let's "cut to the chase" and tell you that those formulas would produce the following results:

degrees of freedom:	208
critical t:	1.96 (in Appendix E)
t-test formula:	0.44 (when you use your calculator!)
Result:	Since the absolute value of 0.44 is less than the critical t of 1.96, we accept the null hypothesis.
Conclusion:	There was no difference between male and female guests last week in their rating of the friendliness of the front-desk clerk at the St. Louis Marriott Hotel.

Now, let's see what happens when you reject the null hypothesis (H_0) and accept the research hypothesis (H_1).

5.1.9.2 Writing the Conclusion of the Two-Group t-Test When You Reject the Null Hypothesis and Accept the Research Hypothesis

Objective: To write the conclusion of the two-group t-test when you have rejected the null hypothesis and accepted the research hypothesis

Let's continue with this same example of the Marriott Hotel, but with the result that we reject the null hypothesis and accept the research hypothesis.

Let's assume that this time you have data on 85 males from last week and their mean score on this question was 7.26 with a standard deviation of 2.35. Let's further suppose that you also have data on 48 females from last week and their mean score on this question was 4.37 with a standard deviation of 3.26, and that you created the table given in Fig. 5.5.

Without going into the details of the formulas for the two-group t-test, these data would produce the following result and conclusion based on Fig. 5.5:

Group	n	Mean	STDEV
1 Males	85	7.26	2.35
2 Females	48	4.37	3.26

Fig. 5.5 Worksheet Data for St. Louis Marriott Hotel for Obtaining a Significant Difference between Males and Females

Null Hypothesis : $\mu_1 = \mu_2$
Research Hypothesis $\mu_1 \neq \mu_2$
degrees of freedom : 131
critical t: 1.96 (in Appendix E)
t-test formula: 5.40 (when you use your calculator!)
Result: Since the absolute value of 5.40 is greater than the critical t of 1.96, we reject the null hypothesis and accept the research hypothesis.

Now, you need to compare the ratings of the men and women to find out which group had the more positive rating of the friendliness of the front-desk clerk using the following rule:

Rule: To summarize the conclusion of the two-group t-test, just compare the means of the two groups, and be sure to use the word "significantly" in your conclusion if you rejected the null hypothesis and accepted the research hypothesis.

A good way to prepare to write the conclusion of the two-group t-test when you are using a rating scale is to place the mean scores of the two groups on a drawing of the scale so that you can visualize the difference of the mean scores. For example, for our Marriott Hotel example above, you would draw this "picture" of the scale in Fig. 5.6:

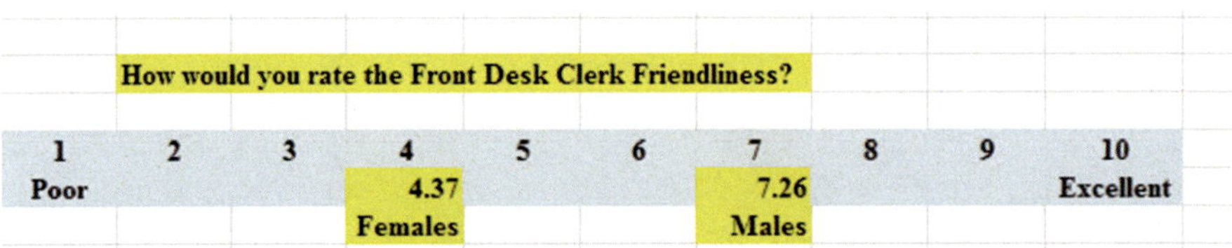

Fig. 5.6 Example of Drawing a "Picture" of the Means of the Two Groups on the Rating Scale

This drawing tells you visually that males had a higher positive rating than females on this item (7.26 vs. 4.37). *And, since you rejected the null hypothesis and accepted the research hypothesis, you know that you have found a significant difference between the two mean scores.*

So, our conclusion needs to contain the following key words:

- Male guests
- Female guests
- Marriott Hotel
- St. Louis
- last week
- significantly
- Front Desk Clerks
- more friendly ***or*** less friendly
- *either*(7.26 vs. 4.37)*or*(4.37 vs. 7.26)

We can use these key words to write the either of two conclusions which are *logically identical*:

Either: Male guests at the Marriott Hotel in St. Louis last week rated the Front Desk Clerks as significantly more friendly than female guests (7.26 vs. 4.37).

Or: Female guests at the Marriott Hotel in St. Louis last week rated the Front Desk Clerks as significantly less friendly than male guests (4.37 vs. 7.26).

Both of these conclusions are accurate, so you can decide which one you want to write. It is your choice.

Also, note that the mean scores in parentheses at the end of these conclusions must match the sequence of the two groups in your conclusion. For example, if you say that: "Male guests rated the Front Desk Clerks as significantly more friendly than female guests," the end of this conclusion should be: (7.26 vs. 4.37) since you mentioned males first and females second.

Alternately, if you wrote that: "Female guests rated the Front Desk Clerks as significantly less friendly than male guests," the end of this conclusion should be: (4.37 vs. 7.26) since you mentioned females first and males second.

Putting the two mean scores at the end of your conclusion saves the reader from having to turn back to the table in your research report to find these mean scores to see how far apart the mean scores were.

If you want to learn more about the two-group t-test, see Crane and Angrosino (1992).

Now, let's discuss FORMULA #1 that deals with the situation in which both groups have more than 30 people in them.

Objective: To use FORMULA #1 for the two-group t-test when both groups have a sample size greater than 30 people

5.2 Formula #1: Both Groups Have More Than 30 People in Them

The first formula we will discuss will be used when you have two groups of people with more than 30 people in each group and one measurement on each person in each group. This formula for the two-group t-test is:

$$t = \frac{\overline{X}_1 - \overline{X}_2}{S_{\overline{X}_1 - \overline{X}_2}} \qquad (5.2)$$

$$\text{where } S_{\overline{X}_1 - \overline{X}_2} = \sqrt{\frac{S_1^2}{n_1} + \frac{S_2^2}{n_2}} \tag{5.3}$$

$$\text{and where degrees of freedom} = df = n_1 + n_2 - 2 \tag{5.1}$$

This formula looks daunting when you first see it, but let's explain some of the parts of this formula:

We have explained the concept of "degrees of freedom" earlier in this chapter, and so you should be able to find the degrees of freedom needed for this formula in order to find the critical value of t in Appendix E.

In the previous chapter, *the formula for the one-group t-test was the following*:

$$t = \frac{\overline{X} - \mu}{S_{\overline{X}}} \tag{4.1}$$

$$\text{where s.e.} = S_{\overline{X}} = \frac{S}{\sqrt{n}} \tag{4.2}$$

For the one-group t-test, you found the mean score and subtracted the population mean from it, and then divided the result by the standard error of the mean (s.e.) to get the result of the t-test. You then compared the t-test result to the critical value of t to see if you either accepted the null hypothesis, or rejected the null hypothesis and accepted the research hypothesis.

The two-group t-test requires a different formula because you have two groups of people, each with a mean score on some variable. You are trying to determine whether to accept the null hypothesis that the *population means of the two groups are equal* (in other words, there is no difference statistically between these two means), or whether the difference between the means of the two groups is "sufficiently large" that you would accept *that there is a significant difference* in the mean scores of the two groups.

The numerator of the two-group t-test asks you to find the difference of the means of the two groups:

$$\overline{X}_1 - \overline{X}_2 \tag{5.4}$$

The next step in the formula for the two-group t-test is to divide the answer you get when you subtract the two means by the standard error of the difference of the two means, and *this is a different standard error of the mean that you found for the one-group t-test because there are two means in the two-group t-test.*

The standard error of the mean when you have two groups of people is called the "standard error of the difference of the means" between the means of the two groups. This formula looks less scary when you break it down into four steps:

1. Square the standard deviation of Group 1, and divide this result by the sample size for Group 1 (n_1).

2. Square the standard deviation of Group 2, and divide this result by the sample size for Group 2 (n_2).
3. Add the results of the above two steps to get a total score.
4. *Take the square root of this total score* to find the standard error of the difference of the means between the two groups, $S_{\overline{X}_1 - \overline{X}_2} = \sqrt{\frac{S_1^2}{n_1} + \frac{S_2^2}{n_2}}$

This last step is the one that gives students the most difficulty when they are finding this standard error using their calculator, because they are in such a hurry to get to the answer that they forget to carry the square root sign down to the last step, and thus get a larger number than they should for the standard error.

5.2.1 An Example of Formula #1 for the Two-Group t-Test

Now, let's use Formula #1 in a situation in which both groups have a sample size greater than 30 people.

Suppose that you have been hired by PepsiCo to do a taste test with teenage boys (ages 13–18) to determine if they like the taste of Pepsi the same as the taste of Coke. The boys are not told the brand name of the soft drink that they taste.

You select a group of boys in this age range, and randomly assign them to one of two groups: (1) Group 1 tastes Coke, and (2) Group 2 tastes Pepsi. Each group rates the taste of their soft drink on a 100-point scale using the following scale in Fig. 5.7:

0	10	20	30	40	50	60	70	80	90	100
poor										excellent

Fig. 5.7 Example of a Rating Scale for a Soft Drink Taste Test (Practical Example)

Suppose you collect these ratings and determine (using your new Excel skills) that the 52 boys in the Coke group had a mean rating of 55 with a standard deviation of 7, while the 57 boys in the Pepsi group had a mean rating of 64 with a standard deviation of 13.

Note that the two-group t-test does not require that both groups have the same sample size. This is another way of saying that the two-group t-test is "robust" (a fancy term that statisticians like to use).

Your data then produce the following table in Fig. 5.8:

Fig. 5.8 Worksheet Data for Soft Drink Taste Test

Group	n	Mean	STDEV
1 Coke	52	55	7
2 Pepsi	57	64	13

Create an Excel spreadsheet, and enter the following information:

B3: Group
B4: 1 Coke
B5: 2 Pepsi
C3: n
D3: Mean
E3: STDEV
C4: 52
D4: 55
E4: 7
C5: 57
D5: 64
E5: 13

Now, widen column B so that it is twice as wide as column A, and center the six numbers and their labels in your table (see Fig. 5.9)

	A	B	C	D	E	F
1						
2						
3		**Group**	**n**	**Mean**	**STDEV**	
4		**1 Coke**	**52**	**55**	**7**	
5		**2 Pepsi**	**57**	**64**	**13**	
6						

Fig. 5.9 Results of Widening Column B and Centering the Numbers in the Cells

B8: Null hypothesis:
B10: Research hypothesis:

Since both groups have a sample size greater than 30, you need to use Formula #1 for the t-test for the difference of the means of the two groups.

Let's "break this formula down into pieces" to reduce the chance of making a mistake.

B13: STDEV1 squared/n1(note that you square the standard deviation of Group 1, and then divide the result by the sample size of Group 1)
B16: STDEV2 squared/n2
B19: D13 + D16
B22: s.e.
B25: critical t
B28: t-test
B31: Result:
B36: Conclusion: (see Fig. 5.10)

Fig. 5.10 Formula Labels for the Two-group t-test

Group	n	Mean	STDEV
1 Coke	52	55	7
2 Pepsi	57	64	13
Null hypothesis:			
Research hypothesis:			
STDEV1 squared / n1			
STDEV2 squared / n2			
D13 + D16			
s.e.			
critical t			
t-test			
Result:			
Conclusion:			

You now need to compute the values of the above formulas in the following cells:

D13: the result of the formula needed to compute cell B13 (use two decimals)
D16: the result of the formula needed to compute cell B16 (use two decimals)
D19: the result of the formula needed to compute cell B19 (use two decimals)
D22: =SQRT(D19) (use two decimals)

This formula should give you a standard error (s.e.) of 1.98.

D25: 1.96
(Since df = n1 + n2 − 2, this gives df = 109 − 2 = 107, and the critical t is, therefore, 1.96 in Appendix E.)
D28: =(D4 – D5)/D22 (use two decimals) (no spaces between symbols)

This formula should give you a value for the t-test of: −4.55.

Nest, check to see if you have rounded off all figures in D13:D28 to two decimal places (see Fig. 5.11).

Fig. 5.11 Results of the t-test Formula for the Soft Drink Taste Test

Group	n	Mean	STDEV
1 Coke	52	55	7
2 Pepsi	57	64	13
Null hypothesis:			
Research hypothesis:			
STDEV1 squared / n1		0.94	
STDEV2 squared / n2		2.96	
D13 + D16		3.91	
s.e.		1.98	
critical t		1.96	
t-test		-4.55	
Result:			
Conclusion:			

Now, write the following sentence in D31 to D34 to summarize the result of the study:

D31: Since the absolute value of −4.55
D32: is greater than the critical t of
D33: 1.96, we reject the null hypothesis
D34: and accept the research hypothesis.

Finally, write the following sentence in D36 to D38 to summarize the conclusion of the study in plain English:

D36: Teenage boys rated the taste of
D37: Pepsi as significantly better than
D38: the taste of Coke (64 vs. 55).

Save your file as: COKE4

Important note: *You are probably wondering why we entered both the result and the conclusion in separate cells instead of in just one cell. This is because if you enter them in one cell, you will be very disappointed when you print out your final spreadsheet, because one of two things will happen that you will not like*: *(1) if you print the spreadsheet to fit onto only one page, the result and the conclusion will force the entire spreadsheet to be printed in such small font size that you will be unable to read it, or (2) if you do not print the final spreadsheet to fit onto one page, both the result and the conclusion will "dribble over" onto a second page instead of fitting the entire spreadsheet onto one page. In either case, your spreadsheet will not have a "professional look."*

Print this file so that it fits onto one page, and write by hand the null hypothesis and the research hypothesis on your printout.

The final spreadsheet appears in Fig. 5.12.

Group	n	Mean	STDEV
1 Coke	52	55	7
2 Pepsi	57	64	13
Null hypothesis:		$\mu_1 = \mu_2$	
Research hypothesis:		$\mu_1 \neq \mu_2$	
STDEV1 squared / n1		0.94	
STDEV2 squared / n2		2.96	
D13 + D16		3.91	
s.e.		1.98	
critical t		1.96	
t-test		-4.55	
Result:		Since the absolute value of - 4.55 is greater than the critical t of 1.96, we reject the null hypothesis and accept the research hypothesis.	
Conclusion:		Teenage boys rated the taste of Pepsi as significantly better than the taste of Coke (64 vs. 55)	

Fig. 5.12 Final Worksheet for the Coke vs. Pepsi Taste Test

Now, let's use the second formula for the two-group t-test which we use whenever either one group, or both groups, have less than 30 people in them.

> Objective: To use Formula #2 for the two-group t-test when one or both groups have less than 30 people in them

Now, let's look at the case when one or both groups have a sample size less than 30 people in them.

5.3 Formula #2: One or Both Groups Have Less Than 30 People in Them

Suppose that a school principal wanted to try out a new method of teaching reading to 4th graders and to compare it to the traditional method of teaching reading in her school. She obtained permission from the school superintendent to do a "pilot test" using two teachers from her school that she considered to be of comparable teaching ability, education, degrees earned, and years of teaching experience. The method of teaching reading that has been used by this school was called the "traditional approach," while the new method of teaching reading was called the "experimental approach." Suppose, further, that these two classes of students had very similar grade equivalent scores in Reading Comprehension at the end of the third grade.

Each of these teachers used just their assigned method of teaching reading, and both teachers taught reading the same amount of time during the school year. At the end of the year, both classes took the Iowa Test of Basic Skills (ITBS) and the grade equivalent scores (GE) in Reading Comprehension were recorded to compare the reading achievement scores of these two classes. For example, a GE score of 4.8 would mean that this pupil was reading at the same developmental level that was typical of a fourth grade pupil in the eighth month of the school year.

Suppose that you have been asked to analyze the data from the Reading Comprehension test scores and to compare the scores of the two classes of pupils using the two-group t-test for independent samples. The hypothetical data is given in Fig. 5.13:

Grade 4 Iowa Tests of Basic Skills: Reading Comprehension
Grade Equivalent (GE) scores

Group	n	Mean	STDEV
1 Traditional approach	26	4.8	0.6
2 Experimental approach	22	5.1	0.4

Fig. 5.13 Worksheet Data for Reading Comprehension Scores (Practical Example)

Null hypothesis : $\mu_1 = \mu_2$
Research hypothesis : $\mu_1 \neq \mu_2$

Note: *Since both groups have a sample size less than 30 people, you need to use Formula #2 in the following steps*:

Create an Excel spreadsheet, and enter the following information:

B1: Grade 4 Iowa Tests of Basic Skills: Reading Comprehension
B2: Grade Equivalent (GE) scores
B4: Group
B5: 1 Traditional approach
B6: 2 Experimental approach
C4: n
D4: Mean
E4: STDEV

Now, widen column B so that it is three times as wide as column A.

To do this, click on B at the top left of your spreadsheet to highlight all of the cells in column B. Then, move the mouse pointer to the right end of the B cell until you get a "cross" sign; then, click on this cross sign and drag the sign to the right until you can read all of the words on your screen. Then, stop clicking!

C5: 26
D5: 4.8
E5: 0.6
C6: 22
D6: 5.1
E6: 0.4

Next, *center the information in cells C4 to E6* by highlighting these cells and then using this step:

Click on the bottom line, second from the left icon, under "Alignment" at the top-center of Home

B9: Null hypothesis:
B11: Research hypothesis: (See Fig. 5.14)

Grade 4 Iowa Tests of Basic Skills: Reading Comprehension
Grade Equivalent (GE) scores

Group	n	Mean	STDEV
1 Traditional approach	26	4.8	0.6
2 Experimental approach	22	5.1	0.4

Null hypothesis:

Research hypothesis:

Fig. 5.14 Reading Comprehension Worksheet Data for Hypothesis Testing

Since both groups have a sample size less than 30, you need to use Formula #2 for the t-test for the difference of the means of two independent samples.

Formula #2 for the two-group t-test is the following:

$$t = \frac{\overline{X}_1 - \overline{X}_2}{S_{\overline{X}_1 - \overline{X}_2}} \tag{5.1}$$

$$\text{where } S_{\overline{X}_1 - \overline{X}_2} = \sqrt{\frac{(n_1 - 1)S_1^2 + (n_2 - 1)S_2^2}{n_1 + n_2 - 2}\left(\frac{1}{n_1} + \frac{1}{n_2}\right)} \tag{5.5}$$

$$\text{and where degrees of freedom} = df = n_1 + n_2 - 2 \tag{5.6}$$

This formula is complicated, and so it will reduce your chance of making a mistake in writing it if you “break it down into pieces” instead of trying to write the formula as one cell entry.

Now, enter these words on your spreadsheet:

B14: (n1 – 1) × STDEV1 squared
B17: (n2 – 1) × STDEV2 squared
B20: $n_1 + n_2 - 2$
B23: $1/n_1 + 1/n_2$
B26: s.e.
B29: critical t:
B32: t-test:
B35: Result:
B40: Conclusion: (see Fig. 5.15)

Grade 4 Iowa Tests of Basic Skills: Reading Comprehension
Grade Equivalent (GE) scores

Group	n	Mean	STDEV
1 Traditional approach	26	4.8	0.6
2 Experimental approach	22	5.1	0.4

Null hypothesis:

Research hypothesis:

(n1 - 1) x STDEV1 squared

(n2 - 1) x STDEV2 squared

n1 + n2 - 2

1/n1 + 1/n2

s.e.

critical t

t-test

Result:

Conclusion:

Fig. 5.15 Reading Comprehension Formula Labels for Two-group t-test

You now need to compute the values of the above formulas in the following cells:

E14: the result of the formula needed to compute cell B14 (use two decimals)
E17: the result of the formula needed to compute cell B17 (use two decimals)
E20: the result of the formula needed to compute cell B20

E23: the result of the formula needed to compute cell B23 (use two decimals)
E26: =SQRT(((E14+E17)/E20)*E23) (no spaces between symbols)

Note the three open-parentheses after SQRT, and the three closed parentheses on the right side of this formula. You need three open parentheses and three closed parentheses in this formula or the formula will not work correctly.

The above formula gives a standard error of the difference of the means equal to 0.15 (two decimals).

E29: enter the critical t value from the t-table in Appendix E in this cell using $df = n_1 + n_2 - 2$ to find the critical t value
E32: =(D5 – D6)/E26 (no spaces between symbols)

Note that you need an open-parenthesis *before D5* and a closed-parenthesis *after D6* so that this answer of −0.30 is *THEN* divided by the standard error of the difference of the means of 0.15, to give a t-test value of −2.00 (note the minus sign here). Use two decimal places for the t-test result (see Fig. 5.16).

Grade 4 Iowa Tests of Basic Skills: Reading Comprehension
Grade Equivalent (GE) scores

Group	n	Mean	STDEV
1 Traditional approach	26	4.8	0.6
2 Experimental approach	22	5.1	0.4

Null hypothesis:

Research hypothesis:

(n1 - 1) x STDEV1 squared	9.00
(n2 - 1) x STDEV2 squared	3.36
n1 + n2 - 2	46
1/n1 + 1/n2	0.08
s.e.	0.15
critical t (df = 46)	1.96
t-test	-2.00

Result:

Conclusion:

Fig. 5.16 Reading Comprehension Two-group t-test Formula Results

Now write the following sentence in D35 to D38 to summarize the *result* of the study:

D35: Since the absolute value of t
D36: of −2.00 is greater than the critical t

D37: of 1.96, we reject the null hypothesis and
D38: accept the research hypothesis.

Finally, write the following sentence in D40 to D42 to summarize the *conclusion* of the study:

D40: The experimental group had significantly
D41: higher grade equivalent (GE) scores than
D42: the traditional group (5.1 vs. 4.8).

Save your file as: Reading3

Print the final spreadsheet so that it fits onto one page.
Write the null hypothesis and the research hypothesis by hand on your printout.
The final spreadsheet appears in Fig. 5.17.

Grade 4 Iowa Tests of Basic Skills: Reading Comprehension
Grade Equivalent (GE) scores

Group	n	Mean	STDEV
1 Traditional approach	26	4.8	0.6
2 Experimental approach	22	5.1	0.4

Null hypothesis:	$\mu_1 = \mu_2$
Research hypothesis:	$\mu_1 \neq \mu_2$
(n1 - 1) x STDEV1 squared	9.00
(n2 - 1) x STDEV2 squared	3.36
n1 + n2 - 2	46
1/n1 + 1/n2	0.08
s.e.	0.15
critical t	1.96
t-test	-2.00
Result:	Since the absolute value of t of - 2.00 is greater than the critical t of 1.96, we reject the null hypothesis and accept the research hypothesis.
Conclusion:	The experimental group had significantly higher grade equivalent (GE) scores than the traditional group (5.1 vs. 4.8).

Fig. 5.17 Reading Comprehension Scores Final Spreadsheet

5.4 End-of-Chapter Practice Problems

1. Howell et al. (2000) studied the effects of a private school education on poor children. Programs were implemented in three U.S. cities that offered children in public schools partial scholarship vouchers to attend private schools. Since more children applied than could be accepted into the program, children were selected randomly (a "true" experiment) to participate in the voucher program. The students who were not selected randomly served as the control group of children. Suppose that you were assigned the task of analyzing the data for the Iowa Test of Basic Skills (ITBS) Vocabulary test that was administered to students in both the experimental group and the control group at the end of the study. Suppose that you created hypothetical dada such that the 52 children in the experimental group had a mean grade-equivalent score of 6.2 with a standard deviation of 0.96, while the 57 children in the control group had a mean grade equivalent score of 5.3 with a standard deviation of 1.12.

   (a) State the null hypothesis and the research hypothesis on an Excel spreadsheet.
   (b) Find the standard error of the difference between the means using Excel
   (c) Find the critical t value using Appendix E, and enter it on your spreadsheet.
   (d) Perform a t-test on these data using Excel. What is the value of t that you obtain?
   Use two decimal places for all figures in the formula section of your spreadsheet.
   (e) State your result on your spreadsheet.
   (f) State your conclusion in plain English on your spreadsheet.
   (g) Save the file as:Voucher2

2. Massachusetts Mutual Financial Group (2010) placed a full-page color ad in *The Wall Street Journal* in which it used a male model hugging a two-year old daughter. The ad had the headline and sub-headline:

What Is the Sign of a Good Decision?

It's knowing your life insurance can help provide income for retirement. And peace of mind until you get there.

Since the majority of the subscribers to *The Wall Street Journal* are men, an interesting research question would be the following:
Research question: "Does a male model in a magazine ad affect adult men's or adult women's willingness to learn more about how life insurance can provide income for retirement?"

Suppose that you have shown one group of adult males (ages 25–39) and one group of adult females (ages 25–39) a mockup of an ad such that both groups saw the ad with a male model. The ads were identical in copy format. The two groups were kept separate during the experiment and could not interact with one another.

At the end of a 1-h discussion of the mockup ad, the respondents were asked the question given in Fig. 5.18:

Item:	"How interested are you in learning more about how life insurance can provide income for retirement?"					
1	2	3	4	5	6	7
Not at all interested						Very Interested

Fig. 5.18 Rating Scale Item for a Magazine Ad Interest Indicator (Practical Example)

The resulting hypothetical data for this question appear in Fig. 5.19:

Item: **"How interested are you in learning more about how life insurance can provide income for retirement?"**

1	2	3	4	5	6	7
Not at all interested						Very Interested

Ad: Male model

Men	Women
5	3
6	4
4	6
7	5
5	2
6	3
5	1
4	3
3	2
6	4
7	3
5	5
6	6
4	3
7	4
5	2
4	5
6	3
3	4
7	5
5	4
6	3
2	2
6	4
1	3
7	5
6	1
5	3
4	2
6	3
5	2
7	5
	3
	4

Fig. 5.19 Worksheet Data for Chap. 5: Practice Problem #2

(a) On your Excel spreadsheet, write the null hypothesis and the research hypothesis.
(b) Create a table that summarizes these data on your spreadsheet and use Excel to find the sample sizes, the means, and the standard deviations of the two groups in this table.
(c) Use Excel to find the standard error of the difference of the means.
(d) Use Excel to perform a two-group t-test. What is the value of t that you obtain (use two decimal places)?
(e) On your spreadsheet, type the *critical value of t* using the t-table in Appendix E.
(f) Type your *result* on the test on your spreadsheet.
(g) Type your *conclusion in plain English* on your spreadsheet.
(h) Save the file as: lifeinsur12

3. Many people in the U.S. feel that the government should make it more difficult to buy a gun, while many others feel that the government should make it easier to buy a gun. Suppose that you were asked to analyze the data from a recent phone survey of registered Republicans and Democrats in Missouri in which Item #13 on the survey asked registered voters how they felt on this issue. You want to test your Excel skills before analyzing the actual data. The hypothetical data for this question are presented in Fig. 5.20:

EFFECT OF POLITICAL PREFERENCE ON ATTITUDE TOWARD GUN CONTROL

Item #13: Buying a gun should be:

1	2	3	4	5
a lot easier than it is now	somewhat easier than it is now	about the same as it is now	somewhat more difficult than it is now	a lot more difficult than it is now

Republicans	Democrats
1	2
2	5
4	4
3	3
5	5
3	4
2	3
4	4
1	5
2	4
3	
2	

Fig. 5.20 Worksheet Data for Chap. 5: Practice Problem #3

(a) State the null hypothesis and the research hypothesis on an Excel spreadsheet.
(b) Find the standard error of the difference between the means using Excel
(c) Find the critical t value using Appendix E, and enter it on your spreadsheet.
(d) Perform a t-test on these data using Excel. What is the value of t that you obtain?
(e) State your result on your spreadsheet.
(f) State your conclusion in plain English on your spreadsheet.
(g) Save the file as: Gun2

References

Crane, J.G. and Angrosino, M.V. Field Projects in Anthropology: A Student Handbook (3rd ed.). Prospect Heights, IL: Waveland Press, Inc., 1992.

Howell, W.G., Wolf, P.J., Campbell, D.E., and Peterson, P.E. "Test-Score Effects of School Vouchers in Dayton, Ohio, New York City, and Washington, D.C.: Evidence from Randomized Field Trials." Paper presented at the annual meeting of the American Political Science Association, Washington, D.C., September 2000.

Keller, G. Statistics for Management and Economics (8th ed.). Mason, OH: South-Western Cengage Learning, 2009.

Mass Mutual Financial Group. What is the Sign of a Good Decision? (Advertisement) *The Wall Street Journal*, September 29, 2010, p. A22.

Zikmund, W.G. and Babin, B.J. Exploring Marketing Research (10th ed.). Mason, OH: South-Western Cengage Learning, 2010.

Chapter 6
Correlation and Simple Linear Regression

There are many different types of "correlation coefficients," but the one we will use in this book is the Pearson product-moment correlation which we will call: *r*.

6.1 What Is a "Correlation?"

Basically, a correlation is a number between –1 and +1 that summarizes the relationship between two variables, which we will call X and Y.

A correlation can be either positive or negative. *A positive correlation means that as X increases, Y increases. A negative correlation means that as X increases, Y decreases.* In statistics books, this part of the relationship is called the *direction* of the relationship (i.e., it is either positive or negative).

The correlation also tells us the *magnitude* of the relationship between X and Y. As the correlation approaches closer to +1, we say that the relationship is *strong and positive.*

As the correlation approaches closer to –1, we say that the relationship is *strong and negative.*

A zero correlation means that there is no relationship between X and Y. This means that neither X nor Y can be used as a predictor of the other.

A good way to understand what a correlation means is to see a "picture" of the scatterplot of points produced in a chart by the data points. Let's suppose that you want to know if variable X can be used to predict variable Y. We will place *the predictor variable X on the x-axis* (the horizontal axis of a chart) and *the criterion variable Yon the y-axis* (the vertical axis of a chart). Suppose, further, that you have collected data given in the scatterplots below (see Figs. 6.1, 6.2, 6.3, 6.4, 6.5, and 6.6).

Figure 6.1 shows the scatterplot for a perfect positive correlation of $r = +1.0$. This means that you can perfectly predict each y-value from each x-value because the data points move "upward-and-to-the-right" along a perfectly-fitting straight line (see Fig. 6.1).

© Springer International Publishing Switzerland 2016
T.J. Quirk, *Excel 2016 for Social Science Statistics*, Excel for Statistics,
DOI 10.1007/978-3-319-39711-5_6

X	Y
1	1
1.5	1.5
2	2
2.5	2.5
3	3
4	4
4.5	4.5
5	5
5.5	5.5
6	6
3.5	3.5

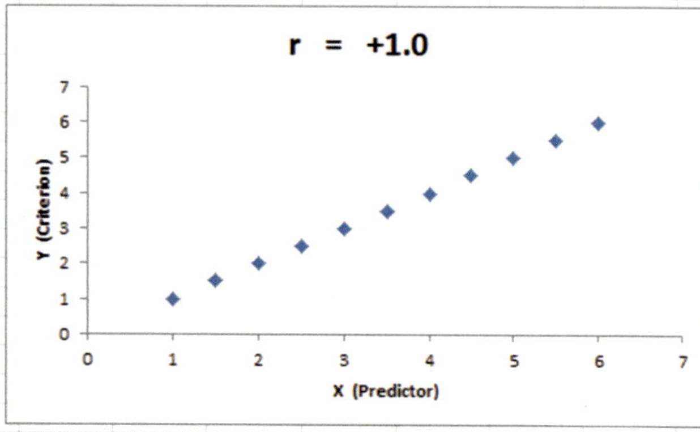

Fig. 6.1 Example of a Scatterplot for a Perfect, Positive Correlation ($r = +1.0$)

Figure 6.2 shows the scatterplot for a moderately positive correlation of $r = +.53$. This means that each x-value can predict each y-value moderately well because you can draw a picture of a "football" around the outside of the data points that move upward-and-to-the-right, but not along a straight line (see Fig. 6.2).

X	Y
1	2
2	4
3	4.5
4	2.5
5	6
6	5
2.5	2
4	5
5.5	3

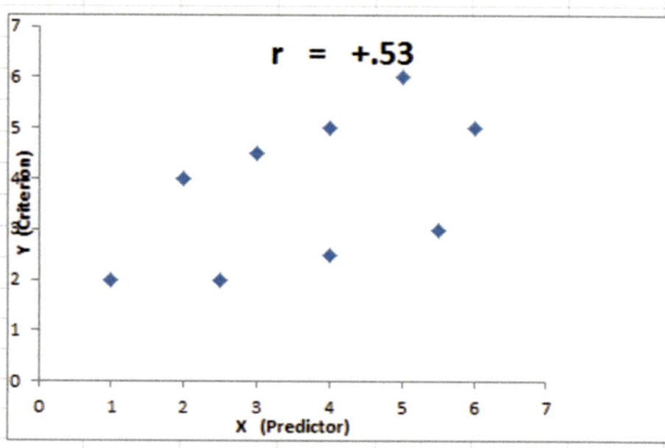

Fig. 6.2 Example of a Scatterplot for a Moderate, Positive Correlation ($r = +.53$)

Figure 6.3 shows the scatterplot for a low, positive correlation of $r = +.23$. This means that each x-value is a poor predictor of each y-value because the "picture" you could draw around the outside of the data points approaches a circle in shape (see Fig. 6.3).

X	Y
1	2
2	4
3	6
4	1
5	5
6	3
3	1
5	2
4.5	5.5
2	1

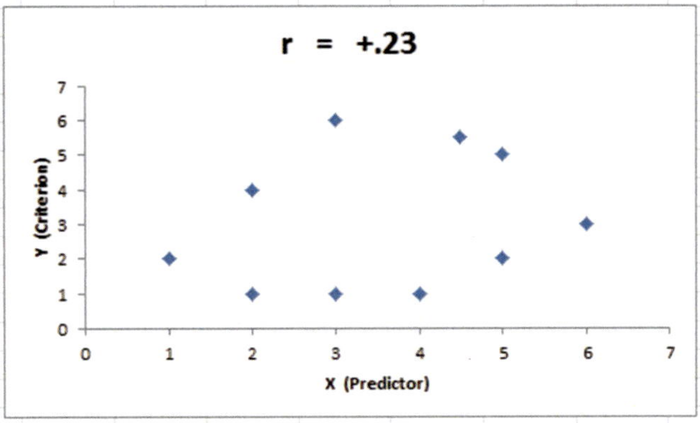

Fig. 6.3 Example of a Scatterplot for a Low, Positive Correlation ($r = +.23$)

We have not shown a Figure of a zero correlation because it is easy to imagine what it looks like as a scatterplot. A zero correlation of $r = .00$ means that there is no relationship between X and Y and the "picture" drawn around the data points would be a perfect circle in shape, indicating that you cannot use X to predict Y because these two variables are not correlated with one another.

Figure 6.4 shows the scatterplot for a low, negative correlation of $r = -.22$ which means that each X is a poor predictor of Y in an inverse relationship, meaning that as X increases, Y decreases (see Fig. 6.4). In this case, it is a negative correlation because the "football" you could draw around the data points slopes down and to the right.

X	Y
1	6
2	5
2	2
3	1
3	5
5	1
6	3
2.5	2.5
3.5	3.4
5.5	5.4

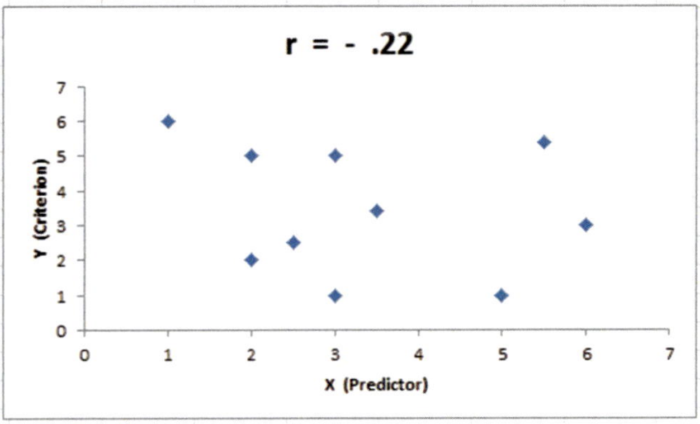

Fig. 6.4 Example of a Scatterplot for a Low, Negative Correlation ($r = -.22$)

Figure 6.5 shows the scatterplot for a moderate, negative correlation of $r = -.39$ which means that X is a moderately good predictor of Y, although there is an inverse relationship between X and Y (i.e., as X increases, Y decreases; see Fig. 6.5). In this case, it is a negative correlation because the "football" you could draw around the data points slopes down and to the right.

X	Y
1	5
2	6
3	1
4	3
5	4
6	2
1.5	3
5	1
2.5	3
4	6

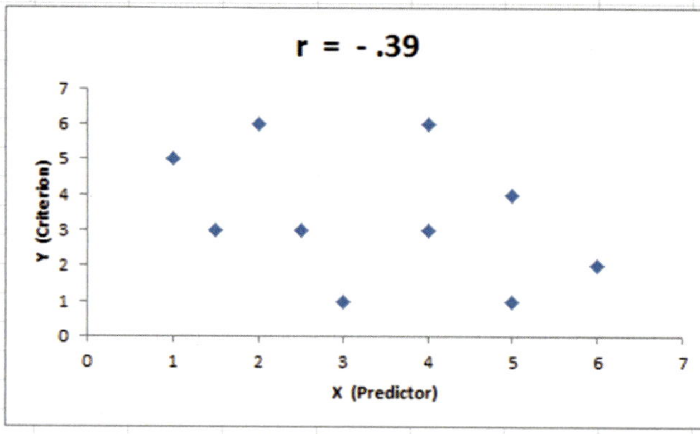

Fig. 6.5 Example of a Scatterplot for a Moderate, Negative Correlation ($r = -.39$)

Figure 6.6 shows a perfect negative correlation of $r = -1.0$ which means that X is a perfect predictor of Y, although in an inverse relationship such that as X increases, Y decreases. The data points fit perfectly along a downward-sloping straight line (see Fig. 6.6).

X	Y
1	6
1.5	5.5
2	5
2.5	4.5
3	4
4	3
5	2
5.5	1.5
6	1
4.5	2.5
3.5	3.5

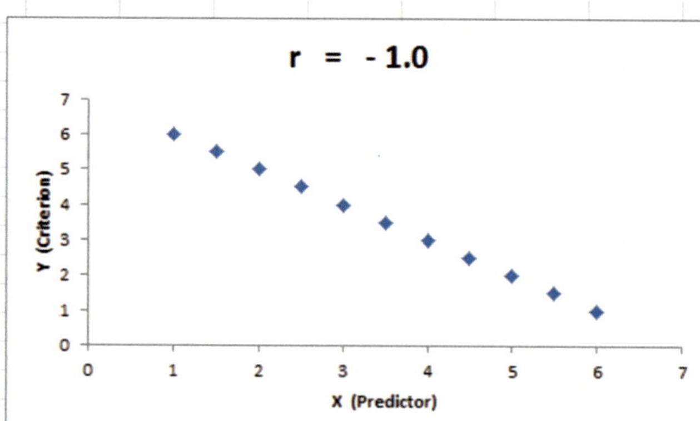

Fig. 6.6 Example of a Scatterplot for a Perfect, Negative Correlation ($r = -1.0$)

Let's explain the formula for computing the correlation *r* so that you can understand where the number summarizing the correlation came from.

In order to help you to understand *where* the correlation number that ranges from −1.0 to +1.0 comes from, we will walk you through the steps involved to use the formula as if you were using a pocket calculator. This is the one time in this book that we will ask you to use your pocket calculator to find a correlation, but knowing how the correlation is computed step-by-step will give you the opportunity to understand *how* the formula works in practice.

To do that, let's create a situation in which you need to find the correlation between two variables.

Suppose that you wanted to find out if there was a relationship between high school grade-point average (HSGPA) and freshman GPA (FRGPA) at a liberal arts college. You have decided to call HSGPA the x-variable (i.e., the predictor variable) and FRGPA as the y-variable (i.e., the criterion variable) in your analysis. To test your Excel skills, you take a random sample of freshmen at the end of their freshman year and record their GPA. The hypothetical data for 8 students appear in Fig. 6.7. (*Note*: *We are using only one decimal place for these GPAs in this example to simplify the mathematical computations*).

Fig. 6.7 Worksheet Data for High School GPA and Frosh GPA (Practical Example)

	X	Y
Student	**High School GPA**	**FROSH GPA**
1	2.8	2.9
2	2.5	2.8
3	3.1	2.8
4	3.5	3.2
5	2.4	2.6
6	2.6	2.3
7	2.4	2.1
8	3.6	3.2
n	8	8
MEAN	2.86	2.74
STDEV	0.48	0.39

Notice also that we have used Excel to find the sample size for both variables, X and Y, and the MEAN and STDEV of both variables. (You can practice your Excel skills by seeing if you get this same result when you create an Excel spreadsheet for these data).

Now, let's use the above table to compute the correlation *r* between HSGPA and FRGPA using your pocket calculator.

6.1.1 Understanding the Formula for Computing a Correlation

Objective: To understand the formula for computing the correlation r

The formula for computing the correlation r is as follows:

$$r = \frac{\frac{1}{n-1}\Sigma\left(X - \overline{X}\right)\left(Y - \overline{Y}\right)}{S_x S_y} \tag{6.1}$$

This formula looks daunting at first glance, but let's "break it down into its steps" to understand how to compute the correlation r.

6.1.2 Understanding the Nine Steps for Computing a Correlation, r

Objective: To understand the nine steps of computing a correlation r

The nine steps are as follows:

Step	Computation	Result
1	Find the sample size n by noting the number of students	8
2	Divide the number 1 by the sample size minus 1 (i.e., 1/7)	0.14286
3	*For each student*, take the HSGPA and subtract the mean HSGPA for the 8 students and call this $X - \overline{X}$ (For example, for student # 6, this would be: 2.6–2.86) *Note*: *With your calculator, this difference is −0.26, but when Excel uses 16 decimal places for every computation, this result could be slightly different for each student*	−0.26
4	*For each student*, take the FRGPA and subtract the mean FRGPA for the 8 students and call this $Y - \overline{Y}$ (For example, for student #6, this would be: 2.3–2.74)	–0.44
5	Then, *for each student*, multiply $(X - \overline{X})$ times $(Y - \overline{Y})$ (For example, for student # 6 this would be: (−0.26) × (−0.44))	+0.1144
6	Add the results of $(X - \overline{X})$ times $(Y - \overline{Y})$ for the eight students	+1.09

Steps 1–6 would produce the Excel table given in Fig. 6.8.

	X	Y			
Student	**High School GPA**	**FROSH GPA**	$X-\bar{X}$	$Y-\bar{Y}$	$(X-\bar{X})(Y-\bar{Y})$
1	2.8	2.9	-0.06	0.16	-0.01
2	2.5	2.8	-0.36	0.06	-0.02
3	3.1	2.8	0.24	0.06	0.01
4	3.5	3.2	0.64	0.46	0.29
5	2.4	2.6	-0.46	-0.14	0.06
6	2.6	2.3	-0.26	-0.44	0.11
7	2.4	2.1	-0.46	-0.64	0.29
8	3.6	3.2	0.74	0.46	0.34
					-------
n	8	8		Total	1.09
MEAN	2.86	2.74			
STDEV	0.48	0.39			

Fig. 6.8 Worksheet for Computing the Correlation, r

Notice that when Excel multiplies a minus number by a minus number, the result is a plus number (for example for student #7: $(-0.46) \times (-0.64) = +0.29$. And when Excel multiplies a minus number by a plus number, the result is a negative number (for example for student #1: $(-0.06) \times (+0.16) = -0.01$.

Note: Excel computes all computation to 16 decimal places. So, when you check your work with a calculator, you frequently get a slightly different answer than Excel's answer.

For example, when you compute above:

$$(X-\overline{X}) \times (Y-\overline{Y}) \text{ for Student \#2, your calculator gives :} \quad (6.2)$$
$$(-0.36) \times (+0.06) = -0.0216$$

As you can see from the table, Excel's answer is -0.02 which is really *more accurate* because Excel uses 16 decimal places for every number, even though only two decimal places are shown in Fig. 6.8.

You should also note that when you do Step 6, you have to be careful to add all of the positive numbers first to get *+1.10* and then add all of the negative numbers second to get *−0.03*, so that when you subtract these two numbers you get *+1.07* as your answer to Step 6. When you do these computations using Excel, this total figure will be *+1.09* because Excel carries every number and computation out to 16 decimal places which is much more accurate than your calculator.

Step		
7	Multiply the answer for step 2 above by the answer for step 6 (0.14286 × 1.09)	0.1557
8	Multiply the STDEV of X times the STDEV of Y (0.48 × 0.39)	0.1872
9	Finally, divide the answer from step 7 by the answer from step 8 (0.1557 divided by 0.1872)	+0.83

This number of *0.83* is the correlation between HSGPA (X) and FRGPA (Y) for these eight students. The number *+0.83* means that there is a strong, positive correlation between these two variables. That is, as HSGPA increases, FRGPA increases. For a more detailed discussion of correlation, see Zikmund and Babin (2010).

You could also use the results of the above table in the formula for computing the correlation r in the following way:

$$\text{correlation r} = \left[(1/(\text{n}-1)) \times \sum \left(X-\overline{X}\right)\left(Y-\overline{Y}\right)\right] / \left(\text{STDEV}_{\text{x}} \times \text{STDEV}_{\text{y}}\right)$$

$$\text{correlation r} = [(1/7) \times 1.09]/[(.48) \times (.39)]$$

$$\text{correlation} = \text{r} = 0.83$$

When you use Excel for these computations, you obtain a slightly different correlation of +0.82 because Excel uses 16 decimal places for all numbers and computations and is, therefore, more accurate than your calculator.

If you want a more detailed explanation of correlation, see Shively (2009), Steinberg (2008), Johnson and Reynolds (2005), and King et al. (1994).

Now, let's discuss how you can use Excel to find the correlation between two variables in a much simpler, and much faster, fashion than using your calculator.

6.2 Using Excel to Compute a Correlation Between Two Variables

Objective: To use Excel to find the correlation between two variables

Suppose that you have been asked to study the relationship between scores on the Law School Admission Test (LSAT) and the GPA of students at the end of their first-year of Law School. The LSAT is a standardized objective measure of Law School applicants and is a required exam for all Law Schools in the U.S. that are approved by the American Bar Association. About 150,000 applicants take this exam every year in the U.S. Because colleges differ in their standards for grades in courses, the LSAT provides a "level playing field" for all applicants by measuring their readiness for Law School in a single examination taken by all the applicants. There are three subtests of the LSAT (Reading Comprehension, Analytical Reasoning, and Logical Reasoning) that produce a single score that ranges between 120 and 180, with an average score about 150.

To test your Excel skills, you take a random sample of students at the end of their first-year of Law School and record their GPA. The hypothetical data appear in Fig. 6.9:

LAW SCHOOL ADMISSION TEST (LSAT)

Is there a relationship between LSAT scores and first-year GPA in law school?

LSAT score	First-year Law School GPA
130	2.65
170	3.72
140	2.85
160	3.25
150	2.75
180	3.95
130	2.35
160	2.74
170	3.65
140	2.55
160	3.72
140	2.35

Fig. 6.9 Worksheet Data for LSAT Scores and GPA (Practical Example)

You want to determine if there is a *relationship* between the LSAT scores and GPA at the end of the first-year of Law School, and you decide to use a correlation to determine this relationship. Let's call the LSAT scores the predictor, X, and first-year GPA the criterion, Y.

Create an Excel spreadsheet with the following information:

A2: LAW SCHOOL ADMISSION TEST (LSAT)
A4: Is there a relationship between LSAT scores and first-year GPA in law school?
B6: LSAT score
C6: First-year Law School GPA
B7: 130

Next, change the width of Columns B and C so that the information fits inside the cells.

Now, complete the remaining figures in the table given above so that B18 is 140 and C18 is 2.35. (Be sure to double-check your figures to make sure that they are correct!) Then, center the information in all of these cells.

A20: n
A21: mean
A22: stdev

Next, define the "name" to the range of data from B7:B18 as: LSAT

We discussed earlier in this book (see Sect. 1.4.4) how to "name a range of data," but here is a reminder of how to do that:

To give a "name" to a range of data:

Click on the top number in the range of data and drag the mouse down to the bottom number of the range.

For example, to give the name: "LSAT" to the cells: B7:B18, click on B7, and drag the pointer down to B18 so that the cells B7:B18 are highlighted on your computer screen. Then, click on:

Formulas
Define name (top center of your screen)
LSAT (in the Name box; see Fig. 6.10)

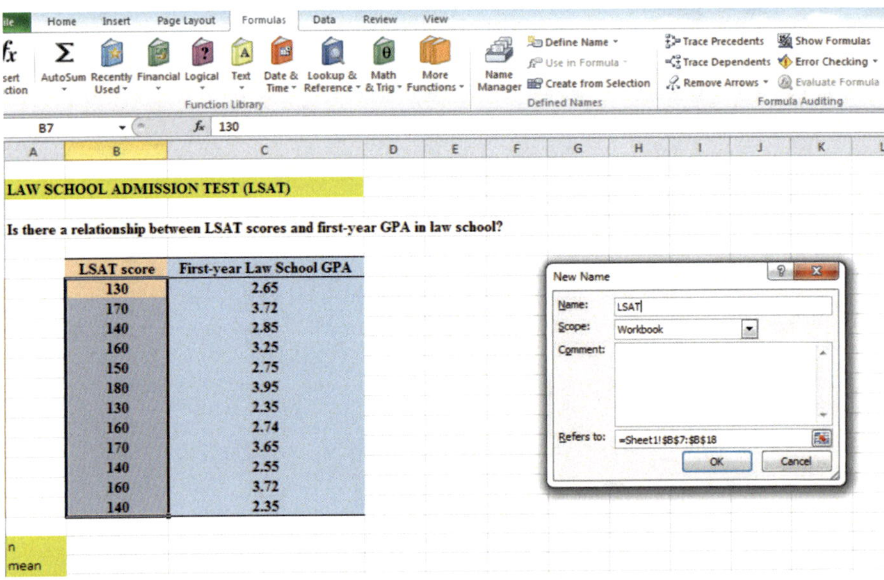

Fig. 6.10 Dialogue Box for Naming a Range of Data as: "LSAT"

OK

Now, repeat these steps to give the name: *GPA* to C7:C18

Finally, click on any blank cell on your spreadsheet to "deselect" cells C7:C18 on your computer screen.

Now, complete the data for these sample sizes, means, and standard deviations in columns B and C so that B22 is 16.58, and C22 is 0.58 (use two decimals for the means and standard deviations; see Fig. 6.11)

	LSAT score	First-year Law School GPA
	130	2.65
	170	3.72
	140	2.85
	160	3.25
	150	2.75
	180	3.95
	130	2.35
	160	2.74
	170	3.65
	140	2.55
	160	3.72
	140	2.35
n	12	12
mean	152.50	3.04
stdev	16.58	0.58

Fig. 6.11 Example of Using Excel to Find the Sample Size, Mean, and STDEV

Objective: Find the correlation between LSAT scores and first-year GPA

B24: correlation
C24: =correl(LSAT,GPA) ; see Fig. 6.12

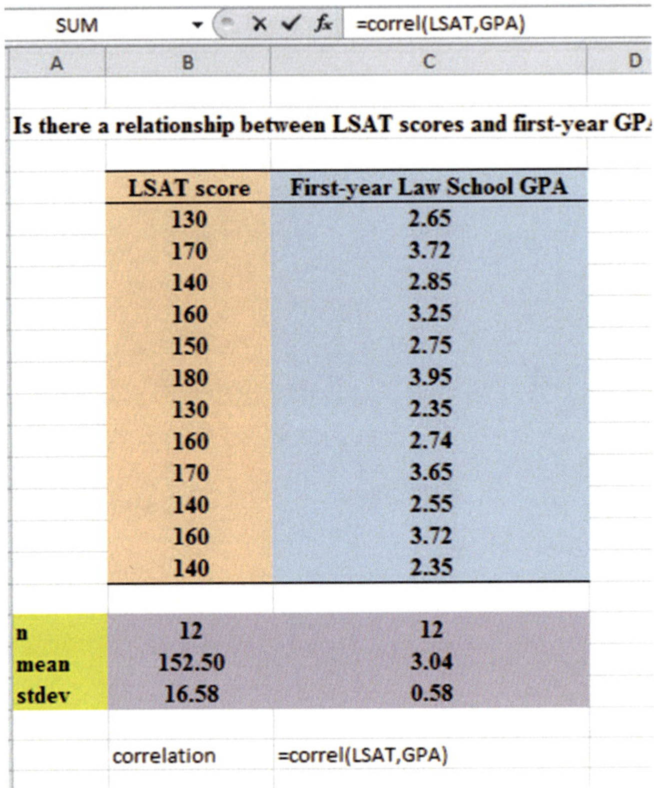
SUM =correl(LSAT,GPA)

Is there a relationship between LSAT scores and first-year GPA

LSAT score	First-year Law School GPA
130	2.65
170	3.72
140	2.85
160	3.25
150	2.75
180	3.95
130	2.35
160	2.74
170	3.65
140	2.55
160	3.72
140	2.35

n	12	12
mean	152.50	3.04
stdev	16.58	0.58
	correlation	=correl(LSAT,GPA)

Fig. 6.12 Example of Using Excel's = correl Function to Compute the Correlation Coefficient

Hit the Enter key to compute the correlation

C24: format this cell to two decimals

Note that the equal sign in =correl(LSAT,GPA) in C24 tells Excel that you are going to use a formula in this cell.

The correlation between LSAT scores (X) and first-year GPA (Y) is +*.89*, a very strong positive correlation. This means that you have evidence that there is a strong relationship between these two variables. In effect, the higher the LSAT score, the higher the first-year GPA in this Law School.

Save this file as: LSAT4

The final spreadsheet appears in Fig. 6.13.

	LSAT score	First-year Law School GPA
	130	2.65
	170	3.72
	140	2.85
	160	3.25
	150	2.75
	180	3.95
	130	2.35
	160	2.74
	170	3.65
	140	2.55
	160	3.72
	140	2.35
n	12	12
mean	152.50	3.04
stdev	16.58	0.58
	correlation	0.89

Fig. 6.13 Final Result of Using the = correl Function to Compute the Correlation Coefficient

6.3 Creating a Chart and Drawing the Regression Line onto the Chart

This section deals with the concept of "linear regression." Technically, the use of a simple linear regression model (i.e., the word "simple" means that only one predictor, X, is used to predict the criterion, Y) requires that the data meet the following four assumptions if that statistical model is to be used:

1. The underlying relationship between the two variables under study (X and Y) is *linear* in the sense that a straight line, and not a curved line, can fit among the data points on the chart.
2. The errors of measurement are independent of each other (e.g. the errors from a specific time period are sometimes correlated with the errors in a previous time period).
3. The errors fit a normal distribution of Y-values at each of the X-values.
4. The variance of the errors is the same for all X-values (i.e., the variability of the Y-values is the same for both low and high values of X).

A detailed explanation of these assumptions is beyond the scope of this book, but the interested reader can find a detailed discussion of these assumptions in Levine et al. (2011, pp. 529–530).

Important note: *Whenever you draw a chart, it is ESSENTIAL that you put the predictor variable (X) on the left, and the criterion variable (Y) on the right in your Excel spreadsheet, so that you know which variable is the predictor variable and which variable is the criterion variable. If you do this, you will save yourself a lot of grief whenever you do a problem involving correlation and simple linear regression using Excel!*

You need to understand that in any chart that has one predictor and a criterion that there are really TWO LINES that can be drawn between the data points:

(1) *One line uses X as the predictor, and Y as the criterion*
(2) *A second line uses Y as the predictor, and X as the criterion*

This means that you have to be very careful to note in your input data the cells that contain X as the predictor, and Y as the criterion. If you get these cells mixed up and reverse them, you will create the wrong line for your data and you will have botched the problem terribly.

This is why we STRONGLY RECOMMEND IN THIS BOOK that you always put the X data (i.e., the predictor variable) on the LEFT of your table, and the Y data (i.e., the criterion variable) on the RIGHT of your table on your spreadsheet so that you don't get these variables mixed up.

Also note that the correlation, r, will be exactly the same correlation no matter which variable you call the predictor variable and which variable you call the criterion variable. The correlation coefficient just summarizes the relationship between two variables, and doesn't care which one is the predictor and which one is the criterion.

Let's suppose that you would like to use LSAT scores as the predictor variable, and that you would like to use it to predict first-year GPA for applicants to this Law School. Since the correlation between these two variables is +.89, this shows that there is a strong, positive relationship and LSAT scores are a good predictor of first-year GPA.

1. Open the file that you saved earlier in this chapter: LSAT4

6.3.1 Using Excel to Create a Chart and the Regression Line Through the Data Points

Objective: To create a chart and the regression line summarizing the relationship between LSAT scores and first-year GPA in Law School

2. Click and drag the mouse to highlight both columns of numbers (B7:C18), *but do not highlight the labels at the top of Column B and Column C.*

Highlight the data set: B7:C18
Insert (top left of screen)
Highlight: Scatter chart icon (immediately above the word: "Charts" at the top center of your screen)
Click on the down arrow on the right of the chart icon
Highlight the top left scatter chart icon (see Fig. 6.14)

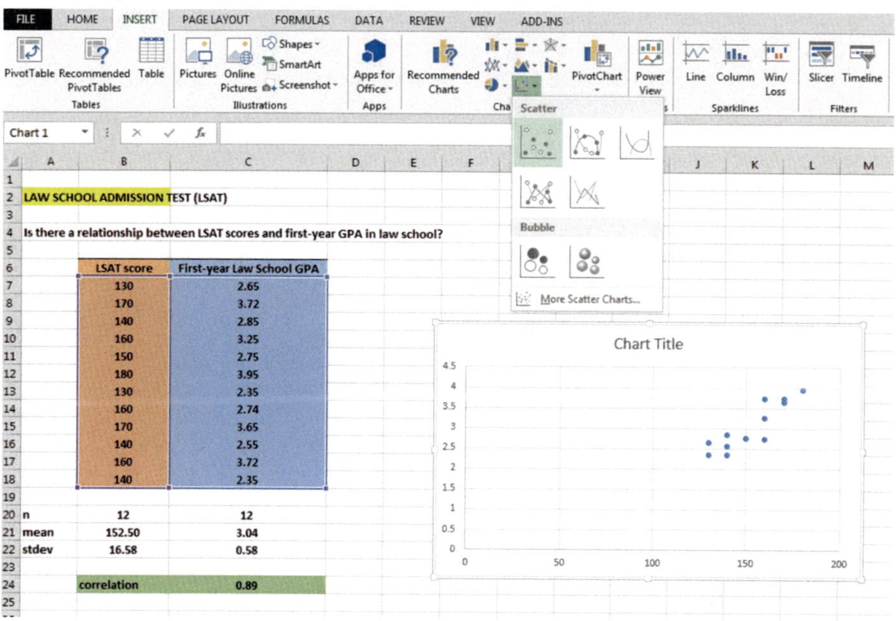

Fig. 6.14 Example of Selecting a Scatter Chart

Click on the top left chart to select it
Click on the "+ icon "to the right of the chart (CHART ELEMENTS).
Click on the check mark next to "Chart Title" **and also** next to "Gridlines" to remove these check marks (see Fig. 6.15)

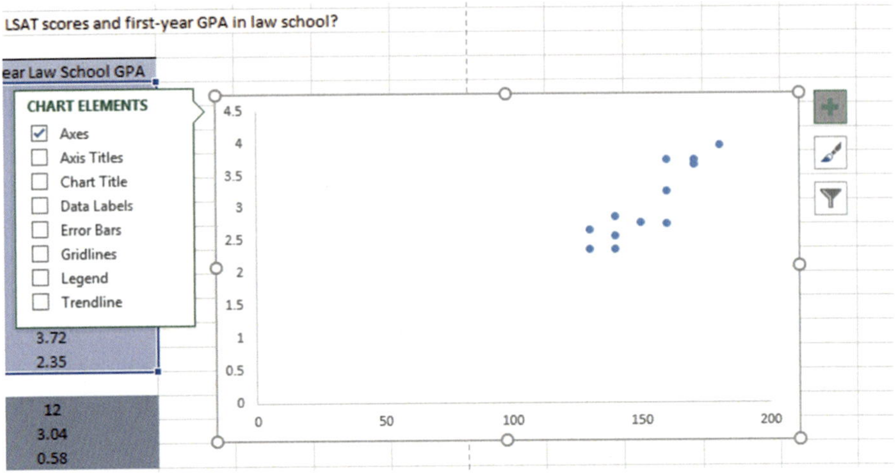

Fig. 6.15 Example of Chart Elements Selected

Click on the box next to: "Chart Title" and then click on the arrow to its right. Then, click on: "Above chart".

Note that the words: "Chart Title" are now in a box at the top of the chart (see Fig. 6.16).

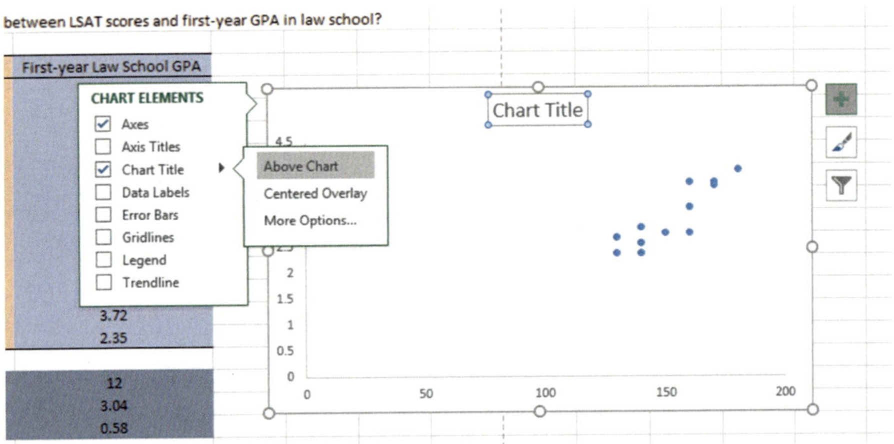

Fig. 6.16 Example of Chart Title Selected

Enter the following Chart Title to the right of $\mathbf{f_x}$ at the top of your screen: RELATIONSHIP BETWEEN LSAT SCORES AND FIRST-YEAR GPA IN LAW SCHOOL (see Fig. 6.17).

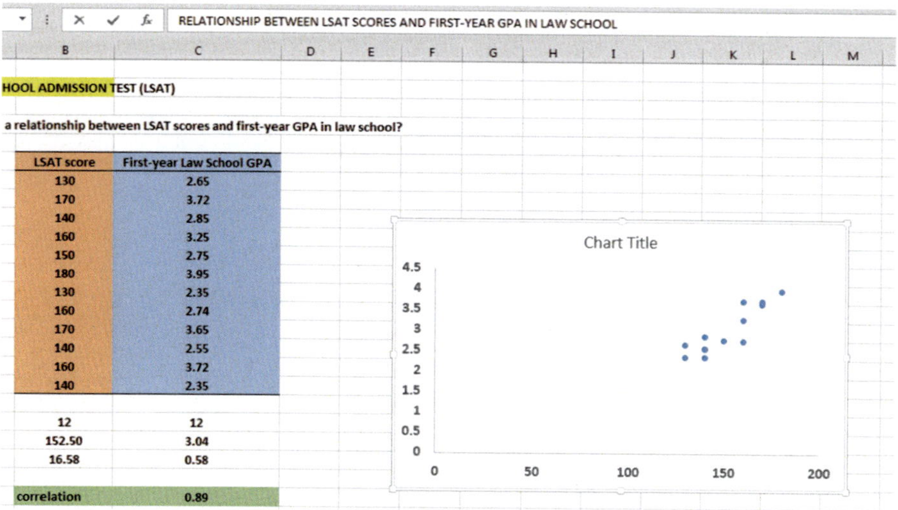

Fig. 6.17 Example of Creating a Chart Title

Hit the Enter Key to enter this chart title onto the chart

Click *inside the chart at the top right corner of the chart* to "deselect" the box around the Chart Title (see Fig. 6.18)

Fig. 6.18 Example of a Chart Title Inserted onto the Chart

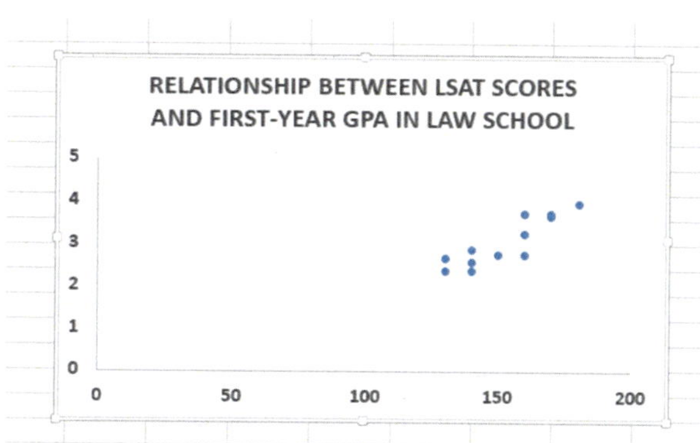

Click on the " + box " to the right of the chart

Add a check mark to the left of "Axis Titles" (This will create an "Axis Title" box on the y-axis of the chart)

Click on the right arrow for: "Axis titles" and then click on: "Primary Horizontal" to remove the check mark in its box (this will create the y-axis title)

Enter the following y-axis title to the right of $\mathbf{f_x}$ at the top of your screen:

First-year GPA in Law School

Then, hit the Enter Key to enter this y-axis title to the chart

Click *inside the chart at the top right corner of the chart* to "deselect" the box around the y-axis title (see Fig. 6.19)

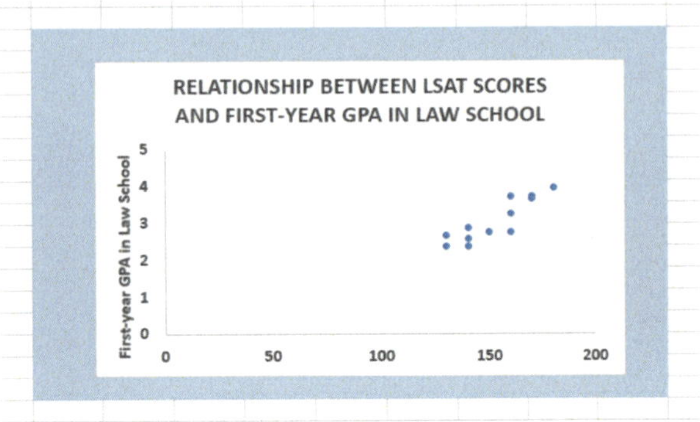

Fig. 6.19 Example of Adding a y-axis Title to the Chart

Click on the " + box " to the right of the chart
Highlight: "Axis Titles" and click on its right arrow
Click on the words: "Primary Horizontal" to add a check mark to its box (this creates an "Axis Title" box on the x-axis of the chart)
Enter the following x-axis title to the right of $\mathbf{f_x}$ at the top of your screen:
LSAT Scores

Then, hit the Enter Key to add this x-axis title to the chart
Click *inside the chart at the top right corner of the chart* to "deselect" the box around the x-axis title (see Fig. 6.20).

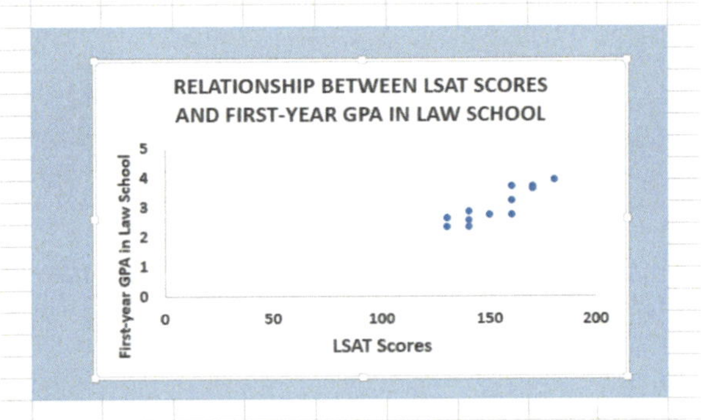

Fig. 6.20 Example of a Chart Title, an x-axis Title, and a y-axis Title

6.3.1.1 Drawing the Regression Line Through the Data Points in the Chart

Objective: To draw the regression line through the data points on the chart

Right-click on any one of the data points inside the chart
Highlight: Add Trendline (see Fig. 6.21)

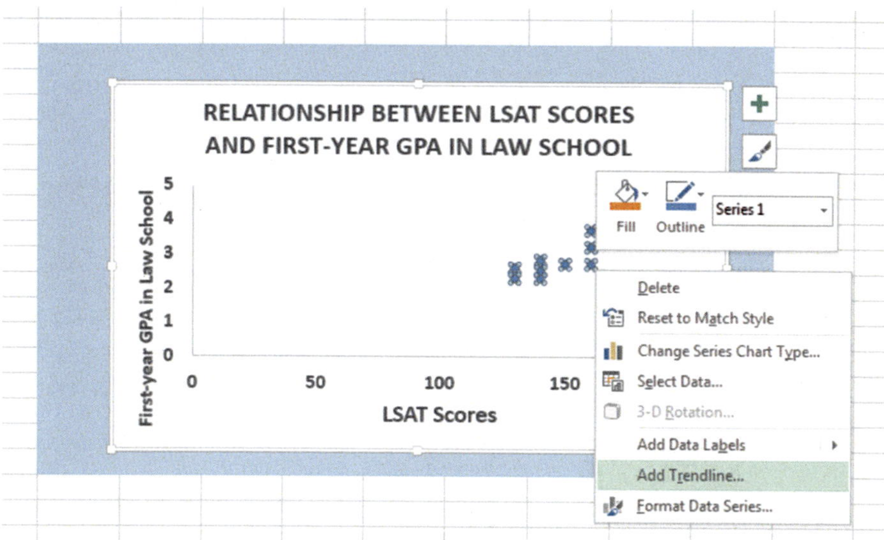

Fig. 6.21 Dialogue Box for Adding a Trendline to the Chart

Click on: Add Trendline

Linear (be sure the "linear" button near the top is selected on the "Format Trendline" dialog box; see Fig. 6.22)

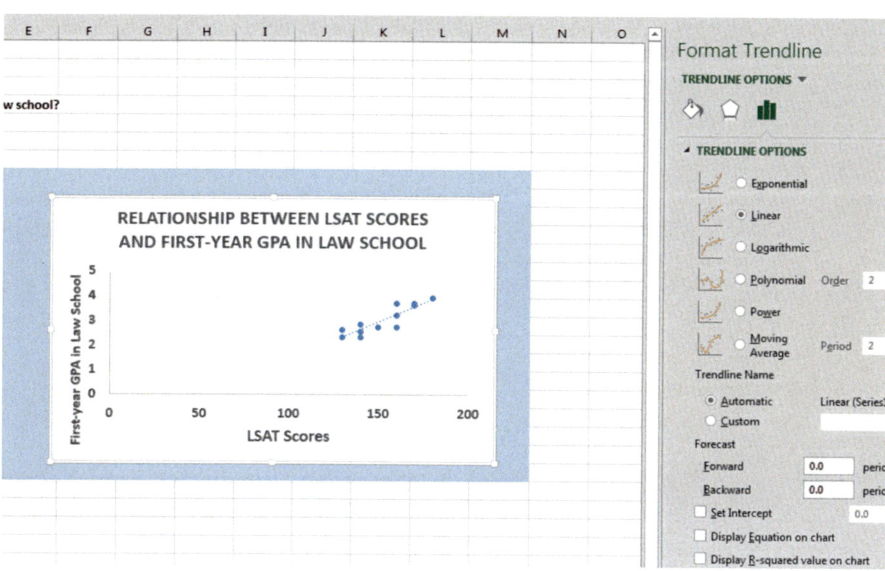

Fig. 6.22 Dialogue Box for a Linear Trendline

Click on the X at the top right of the "Format Trendline" dialog box to close this dialog box

Click on any blank cell *outside the chart* to "deselect" the chart

Save this file as: LSAT5

Your spreadsheet should look like the spreadsheet in Fig. 6.23.

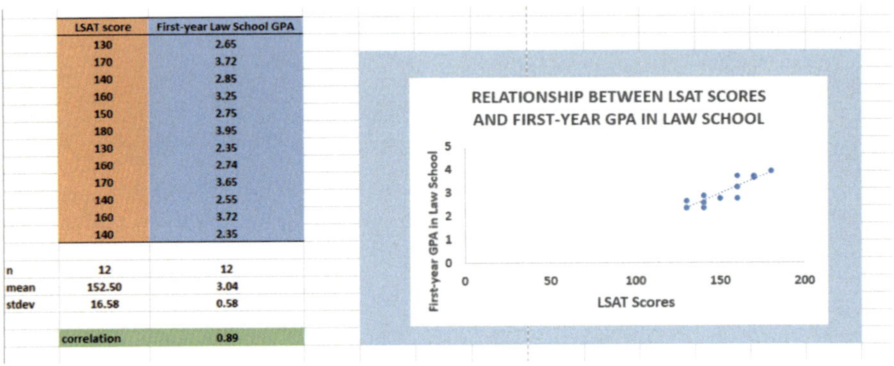

Fig. 6.23 Final Chart with the Trendline Fitted Through the Data Points of the Scatterplot

6.3.1.2 Moving the Chart Below the Table in the Spreadsheet

Objective: To move the chart below the table

Left-click your mouse on *any white space to the right of the top title inside the chart*, keep the left-click down, and drag the chart down and to the left so that the top left corner of the chart is in cell A26, then take your finger off the left-click of the mouse (see Fig. 6.24).

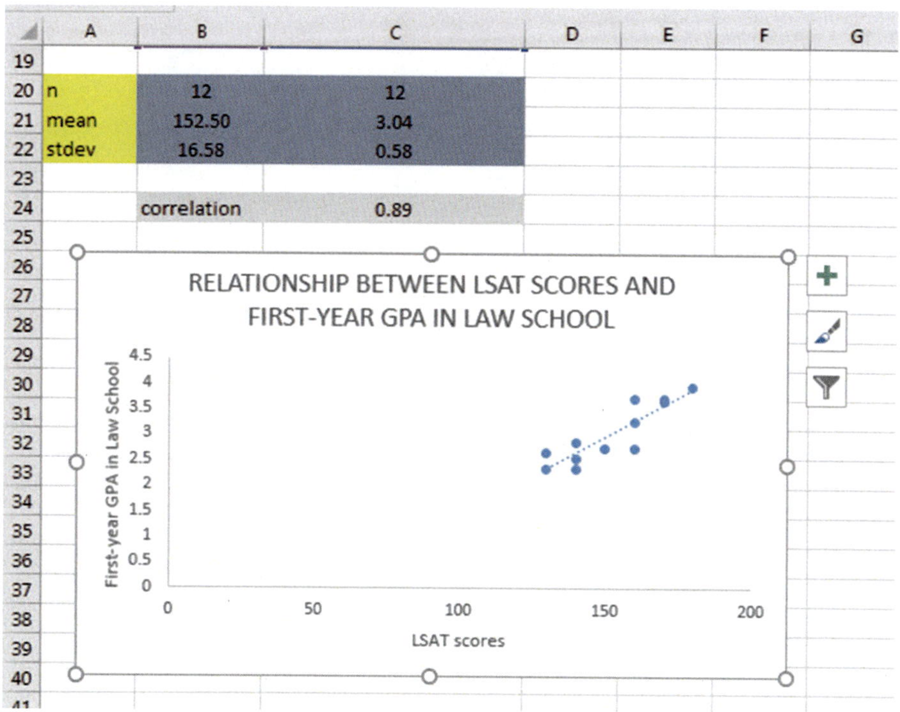

Fig. 6.24 Example of Moving the Chart Below the Table

6.3.1.3 Making the Chart "Longer" So That It Is "Taller"

Objective: To make the chart "longer" so that it is taller

Left-click your mouse on the bottom-center of the chart to create an "up-and-down-arrow" sign, hold the left-click of the mouse down and drag the bottom of the chart down to row 48 to make the chart longer, and then take your finger off the mouse.

6.3.1.4 Making the Chart "Wider"

Objective: To make the chart "wider"

Put the pointer at the middle of the right-border of the chart to create a "left-to-right arrow" sign, and then left-click your mouse and hold the left-click down while you drag the right border of the chart to the middle of Column H to make the chart wider.

Now, click on any blank cell outside the chart to "deselect" the chart (see Fig. 6.25).

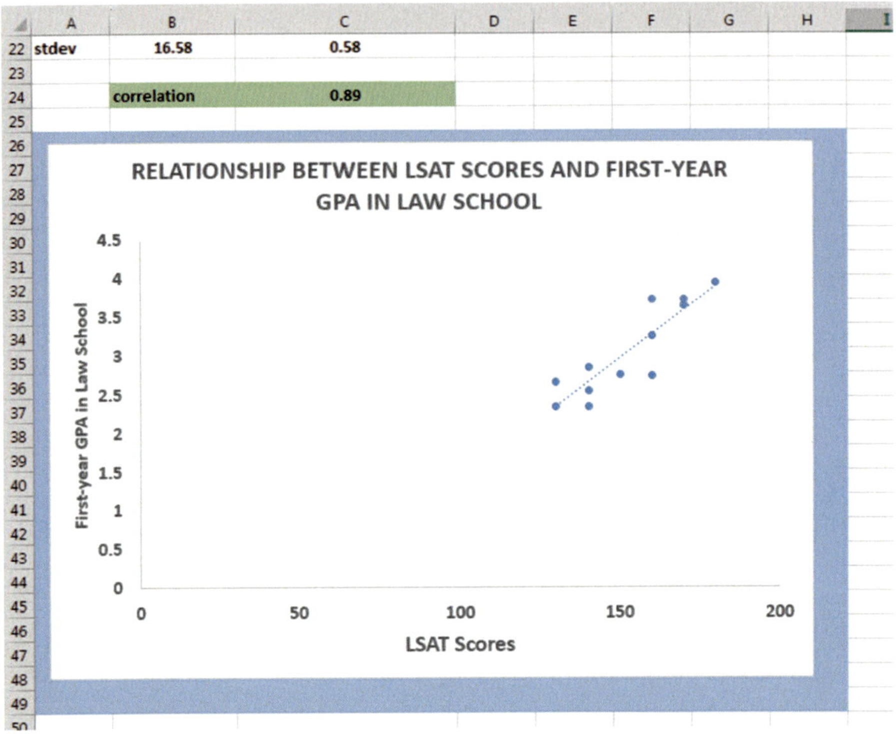

Fig. 6.25 Example of a Chart that is Enlarged to Fit the Cells: A26:H48

Now, click on any blank cell outside the chart to "deselect" the chart
Save this file as: LSAT6

Note: If you printed this spreadsheet now, it is "too big" to fit onto one page, and would "dribble over" onto four pages of printout because the scale needs to be reduced below 100 % in order for this worksheet to fit onto only one page. You need to complete these next steps to print out some, or all, of this spreadsheet.

6.4 Printing a Spreadsheet So That the Table and Chart Fit onto One Page

Objective: To print the spreadsheet so that the table and the chart fit onto one page

Page Layout (top of screen)

Change the scale at the middle icon near the top of the screen "Scale to Fit" by clicking on the down-arrow until it reads "85 %" so that the table and the chart will fit onto one page on your screen (see Fig. 6.26)

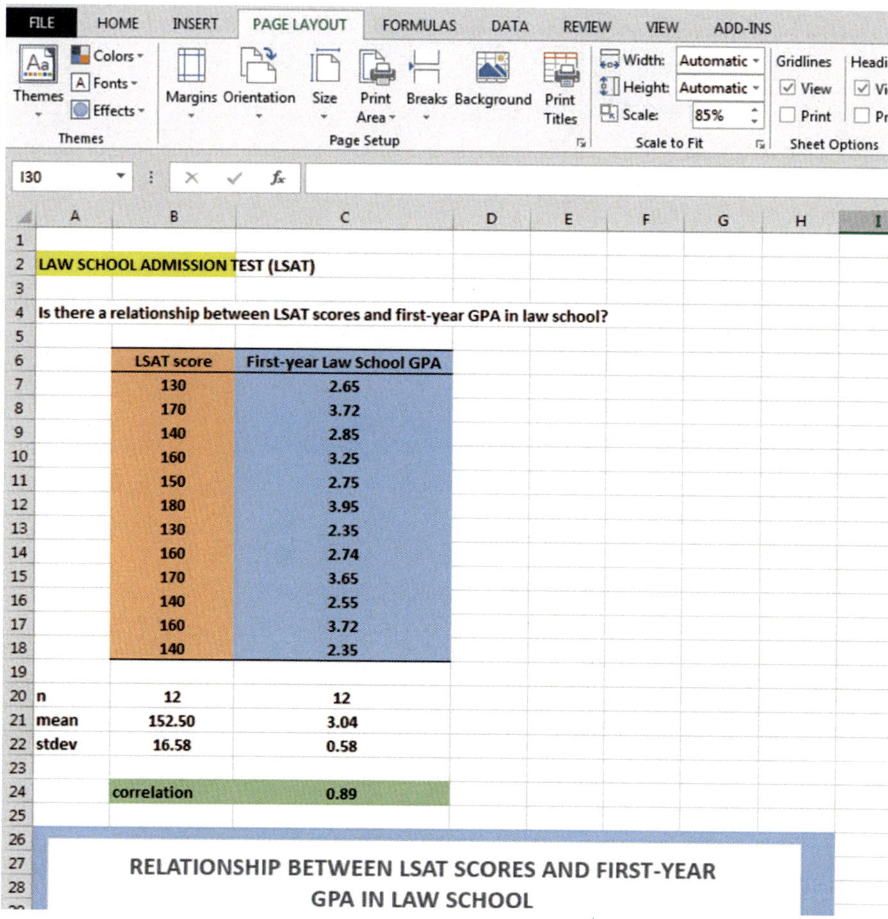

Fig. 6.26 Example of the Page Layout for Reducing the Scale of the Chart to 85 % of Normal Size

File

Print

Print (see Fig. 6.27)

LAW SCHOOL ADMISSION TEST (LSAT)

Is there a relationship between LSAT scores and first-year GPA in law school?

	LSAT score	First-year Law School GPA
	130	2.65
	170	3.72
	140	2.85
	160	3.25
	150	2.75
	180	3.95
	130	2.35
	160	2.74
	170	3.65
	140	2.55
	160	3.72
	140	2.35
n	12	12
mean	152.50	3.04
stdev	16.58	0.58
	correlation	0.89

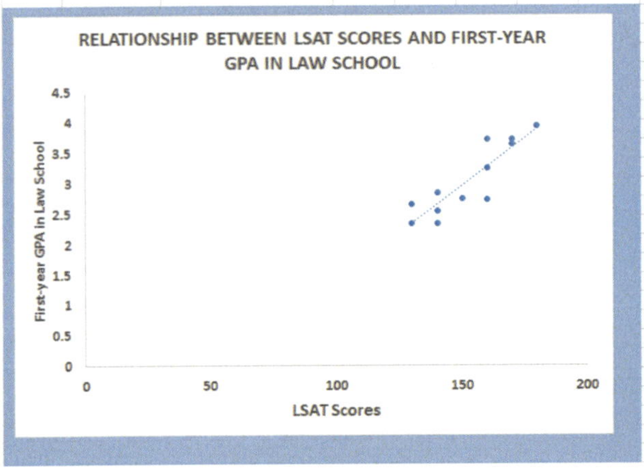

Fig. 6.27 Final Spreadsheet of a Table and a Chart (85 % Scale to Fit Size)

Save your file as: LSAT7

6.5 Finding the Regression Equation

The main reason for charting the relationship between X and Y (i.e., LSAT scores as X and First-year GPA in Law School as Y in our example) is to see if there is a strong relationship between X and Y so that the regression equation that summarizes this relationship can be used to predict Y for a given value of X.

Since we know that the correlation between LSAT scores and First-year GPA is +*.89*, this tells us that it makes sense to use LSAT scores to predict First-year GPA in Law School based on our past data from this Law School.

We now need to find that regression equation that is the equation of the "best-fitting straight line" through the data points.

Objective: To find the regression equation summarizing the relationship between X and Y.

In order to find this equation, we need to check to see if your version of Excel contains the "Data Analysis ToolPak" necessary to run a regression analysis.

6.5.1 *Installing the Data Analysis ToolPak into Excel*

Objective: To install the Data Analysis ToolPak into Excel

Since there are currently four versions of Excel in the marketplace (2007, 2010, 2013, 2016), we will give a brief explanation of how to install the Data Analysis ToolPak into each of these versions of Excel.

6.5.1.1 Installing the Data Analysis ToolPak into Excel 2016

Open a new Excel spreadsheet

Click on: Data (at the top of your screen)

Look at the top of your monitor screen. Do you see the words: "Data Analysis" at the far right of the screen? If you do, the Data Analysis ToolPak for Excel 2016 was correctly installed when you installed Office 2016, and you should skip ahead to Sect. 6.5.2.

If the words: "Data Analysis" are not at the top right of your monitor screen, then the ToolPak component of Excel 2016 was not installed when you installed Office 2016 onto your computer. If this happens, you need to follow these steps:

File
Options (bottom left of screen)

Note: This creates a dialog box with "Excel Options" at the top left of the box)

Add-Ins (on left of screen)
Manage: Excel Add-Ins (at the bottom of the dialog box)
Go (at bottom center of dialog box)
Highlight: Analysis ToolPak (in the Add-Ins dialog box)
Put a check mark to the left of Analysis Toolpak
OK (at the right of this dialog box)
Data

You now should have the words: "Data Analysis" at the top right of your screen to show that this feature has been installed correctly

Note: If these steps do not work, you should try these steps instead:

File/Options (bottom left)/Add-ins/Analysis ToolPak/Go/click to the left of Analysis ToolPak to add a check mark/OK

If you need help doing this, ask your favorite "computer techie" for help.

You are now ready to skip ahead to Sect. 6.5.2

6.5.1.2 Installing the Data Analysis ToolPak into Excel 2013

Open a new Excel spreadsheet

Click on: Data (at the top of your screen)

Look at the top of your monitor screen. Do you see the words: "Data Analysis" at the far right of the screen? If you do, the Data Analysis ToolPak for Excel 2013 was correctly installed when you installed Office 2013, and you should skip ahead to Sect. 6.5.2.

If the words: "Data Analysis" are not at the top right of your monitor screen, then the ToolPak component of Excel 2013 was not installed when you installed Office 2013 onto your computer. If this happens, you need to follow these steps:

File
Options (bottom left of screen)

Note: This creates a dialog box with "Excel Options" at the top left of the box

Add-Ins (on left of screen)
Manage: Excel Add-Ins (at the bottom of the dialog box)
Go (at bottom center of dialog box)
Highlight: Analysis ToolPak (in the Add-Ins dialog box)
Put a check mark to the left of Analysis Toolpak
OK (at the right of this dialog box)
Data

You now should have the words: "Data Analysis" at the top right of your screen to show that this feature has been installed correctly If you get a prompt asking you for the "installation CD," put this CD in the CD drive and click on: OK

Note: If these steps do not work, you should try these steps instead:

File/Options (bottom left)/Add-ins/Analysis ToolPak/Go/click to the left of Analysis ToolPak to add a check mark/OK

If you need help doing this, ask your favorite "computer techie" for help.

You are now ready to skip ahead to Sect. 6.5.2

6.5.1.3 Installing the Data Analysis ToolPak into Excel 2010

Open a new Excel spreadsheet

Click on: Data (at the top of your screen)

Look at the top of your monitor screen. Do you see the words: "Data Analysis" at the far right of the screen? If you do, the Data Analysis ToolPak for Excel 2010 was correctly installed when you installed Office 2010, and you should skip ahead to Sect. 6.5.2.

If the words: "Data Analysis" are not at the top right of your monitor screen, then the ToolPak component of Excel 2010 was not installed when you installed Office 2010 onto your computer. If this happens, you need to follow these steps:

File
Options
Excel options (creates a dialog box)
Add-Ins
Manage: Excel Add-Ins (at the bottom of the dialog box)
Go
Highlight: Analysis ToolPak (in the Add-Ins dialog box)
OK
Data (You now should have the words: "Data Analysis" at the top right of your screen)

If you get a prompt asking you for the "installation CD," put this CD in the CD drive and click on: OK

Note: If these steps do not work, you should try these steps instead:

File/Options (bottom left)/Add-ins/Analysis ToolPak/Go/click to the left of Analysis ToolPak to add a check mark/OK

If you need help doing this, ask your favorite "computer techie" for help.

You are now ready to skip ahead to Sect. 6.5.2.

6.5.1.4 Installing the Data Analysis ToolPak into Excel 2007

Open a new Excel spreadsheet

Click on: Data (at the top of your screen

If the words "Data Analysis" do not appear at the top right of your screen, you need to install the Data Analysis ToolPak using the following steps:

Microsoft Office button (top left of your screen)
Excel options (bottom of dialog box)
Add-ins (far left of dialog box)
Go (to create a dialog box for Add-Ins)

Highlight: Analysis ToolPak
OK (If Excel asks you for permission to proceed, click on: Yes)
Data (You should now have the words: "Data Analysis" at the top right of your screen)

If you need help doing this, ask your favorite "computer techie" for help.
You are now ready to skip ahead to Sect. 6.5.2.

6.5.2 Using Excel to Find the SUMMARY OUTPUT of Regression

You have now installed *ToolPak*, and you are ready to find the regression equation for the "best-fitting straight line" through the data points by using the following steps:

Open the Excel file: *LSAT7*(if it is not already open on your screen)

Note: *If this file is already open, and there is a border around the chart, you need to click on any empty cell outside of the chart to deselect the chart.*

Now that you have installed *Toolpak*, you are ready to find the regression equation summarizing the relationship between LSAT scores and First-year GPA in Law School in your data set.

Remember that you gave the name: ***LSAT*** to the X data (the predictor), and the name: **GPA** to the Y data (the criterion) in a previous section of this chapter (see Sect. 6.2)

Data (top of screen)
Data analysis (far right at top of screen; see Fig. 6.28)

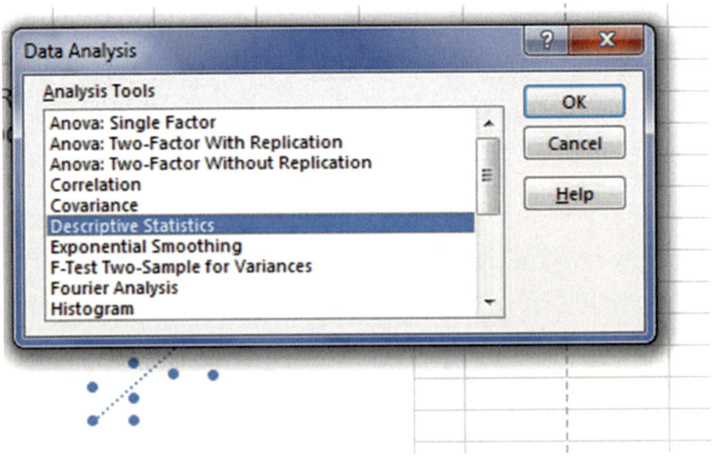

Fig. 6.28 Example of Using the Data/Data Analysis Function of Excel

Scroll down the dialog box using the down arrow and highlight: Regression (see Fig. 6.29)

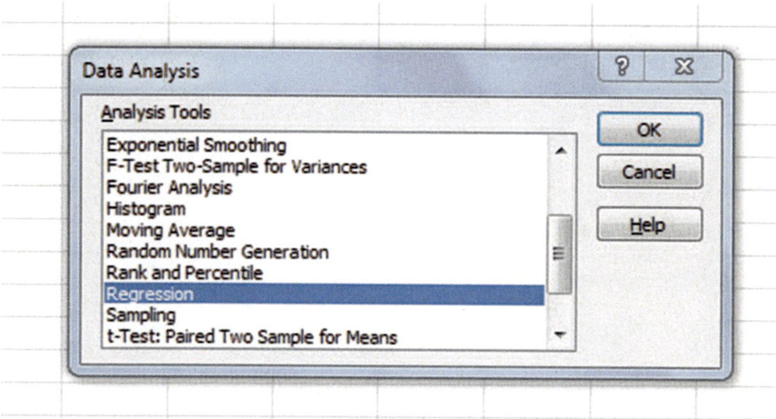

Fig. 6.29 Dialogue Box for Creating the Regression Function in Excel

OK

Input Y Range: GPA
Input X Range: LSAT

Click on the "button" to the left of Output Range to select this, and enter A50 in the box as the place on your spreadsheet to insert the Regression analysis in cell A50

OK

The *SUMMARY OUTPUT* should now be in cells: A50:I67

Widen Column A so that all of the words in the SUMMARY OUTPUT are readable.

Now, change the data in the following three cells to Number format (two decimal places) by first clicking on "Home" at the top left of your screen:

B53
B66
B67

Now, change the format for all other numbers that are in decimal format to number format, three decimal places.

Next, widen all columns so that all of the labels fit inside the column widths.

Then, center all numbers in their cells.

Print the file so that it fits onto one page. (*Hint*: *Change the scale under "Page Layout" to 65 % to make it fit.*) Your file should be like the file in Fig. 6.30.

LAW SCHOOL ADMISSION TEST (LSAT)

Is there a relationship between LSAT scores and first-year GPA in law school?

	LSAT score	First-year Law School GPA
	130	2.65
	170	3.72
	140	2.85
	160	3.25
	150	2.75
	180	3.95
	130	2.35
	160	2.74
	170	3.65
	140	2.55
	160	3.72
	140	2.35
n	12	12
mean	152.50	3.04
stdev	16.58	0.58
	correlation	0.89

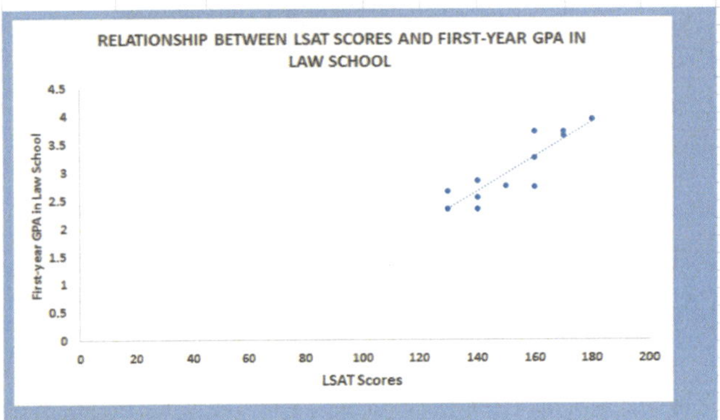

SUMMARY OUTPUT

Regression Statistics	
Multiple R	0.89
R Square	0.787
Adjusted R Square	0.766
Standard Error	0.281
Observations	12

ANOVA

	df	SS	MS	F	Significance F
Regression	1	2.932	2.932	37.037	0.000
Residual	10	0.792	0.079		
Total	11	3.723			

	Coefficients	Standard Error	t Stat	P-value	Lower 95%	Upper 95%	Lower 95.0%	Upper 95.0%
Intercept	-1.70	0.784	-2.172	0.055	-3.451	0.044	-3.451	0.044
X Variable 1	0.03	0.005	6.086	0.000	0.020	0.043	0.020	0.043

Fig. 6.30 Final Spreadsheet of Correlation and Simple Linear Regression including the SUMMARY OUTPUT for the Data

Save the resulting file as: LSAT8

Note the following problem with the summary output. Whoever wrote the computer program for this version of Excel made a mistake and gave the name: "Multiple R" to cell A53.

This is not correct. Instead, cell A53 should say: "correlation r" since this is the notation that we are using for the correlation between X and Y.

You can now use your printout of the regression analysis to find the regression equation that is the best-fitting straight line through the data points.

But first, let's review some basic terms.

6.5.2.1 Finding the y-intercept, a, of the Regression Line

The point on the y-axis that the regression line would intersect the y-axis if it were extended to reach the y-axis is called the "y-intercept" and *we will use the letter "a" to stand for the y-intercept of the regression line.* The y-intercept on the SUMMARY OUTPUT on the previous page is *−1.70 and appears in cell B66* (note the minus sign). This means that if you were to draw an imaginary line continuing down the regression line toward the y-axis that this imaginary line would cross the y-axis at −1.70. This is why *a* is called the "y-intercept."

6.5.2.2 Finding the Slope, b, of the Regression Line

The "tilt" of the regression line is called the "slope" of the regression line. It summarizes to what degree the regression line is either above or below a horizontal line through the data points. If the correlation between X and Y were zero, the regression line would be exactly horizontal to the X-axis and would have a zero slope.

If the correlation between X and Y is positive, the regression line would "slope upward to the right" above the X-axis. Since the regression line in Fig. 6.30 slopes upward to the right, the slope of the regression line is *+0.03* as given in cell *B67. We will use the notation "b" to stand for the slope of the regression line.* (Note that Excel calls the slope of the line: "X Variable 1" in the Excel printout.)

Since the correlation between LSAT scores and First-year GPA was +*.89*, you can see that the regression line for these data "slopes upward to the right" through the data. Note that the SUMMARY OUTPUT of the regression line in Fig. 6.30 gives a correlation, r, of +*.89* in cell *B53*.

If the correlation between X and Y were negative, the regression line would "slope down to the right" above the X-axis. This would happen whenever the correlation between X and Y is a negative correlation that is between zero and minus one (0 and −1).

6.5.3 Finding the Equation for the Regression Line

To find the regression equation for the straight line that can be used to predict First-year GPA from LSAT scores, we only need two numbers in the SUMMARY OUTPUT in Fig. 6.30: *B66 and B67*.

$$\text{The format for the regression line is}: \quad Y = a + bX \tag{6.3}$$

where a = *the y-intercept* (−1.70 in our example in cell B66)
and b = *the slope of the line* (+0.03 in our example in cell B67)
Therefore, the equation for the best-fitting regression line for our example is:

$$\mathrm{Y = a + bX}$$

$$\boxed{\mathrm{Y} = -1.70 + 0.03\,\mathrm{X}}$$

Remember that Y is the First-year GPA that we are trying to predict, using the LSAT scores as the predictor, X.

Let's try an example using this formula to predict the First-year GPA for a hypothetical student.

6.5.4 Using the Regression Line to Predict the y-Value for a Given x-Value

Objective: Find the First-year GPA from an LSAT score of 150.

Since the LSAT score is 150 (i.e., X = 150), substituting this number into our regression equation gives:

$$\mathrm{Y} = -1.70 + 0.03(150)$$
$$\mathrm{Y} = -1.70 + 4.50$$
$$\mathrm{Y} = 2.80$$

Important note: If you look at your chart, if you go directly upwards from an LSAT score of 150 until you hit the regression line, you see that you hit this line just under the number 3on the y-axis to the left (actually, it is 2.80), the result above for predicting First-year GPA from an LSAT score of 150.

Now, let's do a second example and predict what the First-year GPA would be if we used an LSAT score of 170.

$$\begin{aligned}Y &= -1.70 + 0.03X \\ Y &= -1.70 + 0.03(170) \\ Y &= -1.70 + 5.10 \\ Y &= 3.40\end{aligned}$$

Important note: If you look at your chart, if you go directly upwards from an LSAT score of 170 until you hit the regression line, you see that you hit this line just under the number 3.5 on the y-axis to the left (actually it is 3.40), the result above for predicting First-year GPA from an LSAT score of 170.

For a more detailed discussion of regression, see Black (2010).

6.6 Adding the Regression Equation to the Chart

Objective: To Add the Regression Equation to the Chart

If you want to include the regression equation within the chart next to the regression line, you can do that, but a word of caution first.

Throughout this book, we are using the regression equation for one predictor and one criterion to be the following:

$$Y = a + bX \tag{6.3}$$

where a = y-intercept and
b = slope of the line

See, for example, the regression equation in Sect. 6.5.3 where the y-intercept was $a = -1.70$ and the slope of the line was $b = +0.03$ to generate the following regression equation:

$$Y = -1.70 + 0.03X$$

However, Excel 2016 uses a slightly different regression equation (which is logically identical to the one used in this book) when you add a regression equation to a chart:

$$Y = bX + a \tag{6.4}$$

where a = y-intercept and
b = slope of the line

Note that this equation is identical to the one we are using in this book with the terms arranged in a different sequence.

For the example we used in Sect. 6.5.3, Excel 2016 would write the regression equation on the chart as:

$$Y = 0.03\,X - 1.70$$

This is the format that will result when you add the regression equation to the chart using Excel 2016 using the following steps:

Open the file: *LSAT8* (*that you saved in Sect. 6.5.2*)

Click just *inside* the outer border of the chart in the top right corner to add the "border" around the chart in order to "select the chart" for changes you are about to make

Right-click on any of the data-points in the chart

Highlight: Format Trendline, and click on it to select this command

The "Linear button" near the top of the dialog box will already be selected (on its left)

Scroll down this dialog box and click on: Display Equation on chart (near the bottom of the dialog box; see Fig. 6.31)

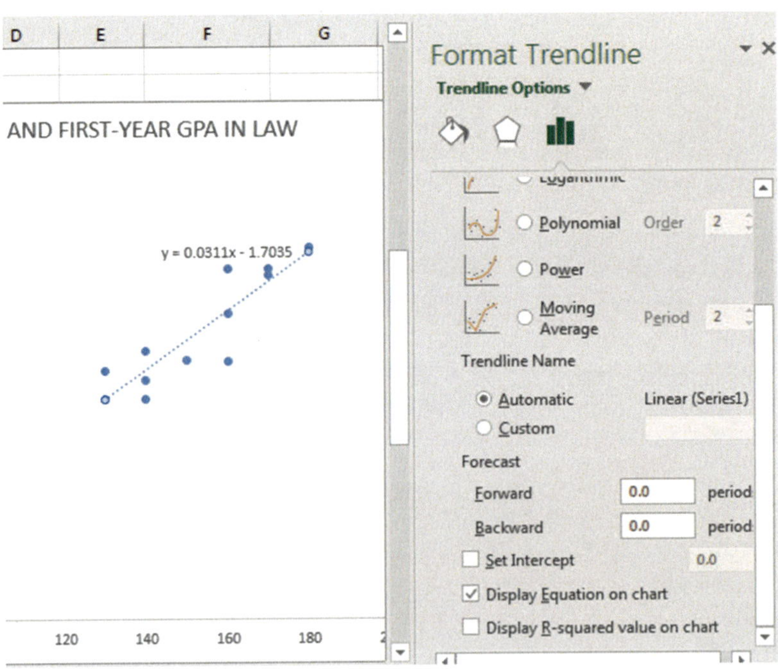

Fig. 6.31 Dialogue Box for Adding the Regression Equation to the Chart Next to the Regression Line on the Chart

Click on the X at the top right of the Format Trendline dialogue box to remove this box.
Click on any empty cell outside of the chart to deselect the chart.

Note that the regression equation on the chart is in the following form next to the regression line on the chart (see Fig. 6.32).

LAW SCHOOL ADMISSION TEST (LSAT)

Is there a relationship between LSAT scores and first-year GPA in law school?

	LSAT score	First-year Law School GPA
	130	2.65
	170	3.72
	140	2.85
	160	3.25
	150	2.75
	180	3.95
	130	2.35
	160	2.74
	170	3.65
	140	2.55
	160	3.72
	140	2.35
n	12	12
mean	152.50	3.04
stdev	16.58	0.58
	correlation	0.89

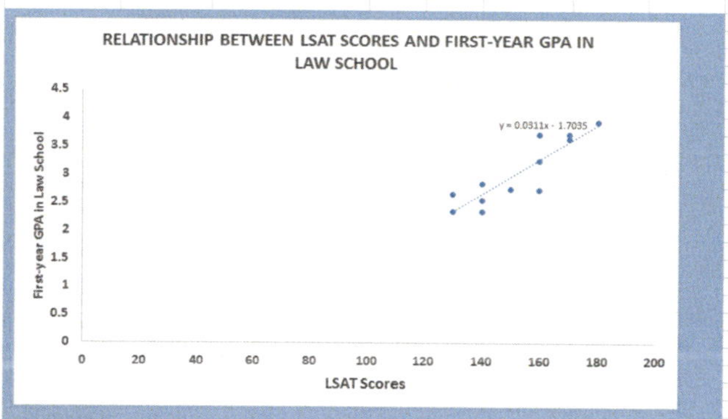

SUMMARY OUTPUT

Regression Statistics	
Multiple R	0.89
R Square	0.787
Adjusted R Square	0.766
Standard Error	0.281
Observations	12

ANOVA

	df	*SS*	*MS*	*F*	*Significance F*
Regression	1	2.932	2.932	37.037	0.000
Residual	10	0.792	0.079		
Total	11	3.723			

	Coefficients	*Standard Error*	*t Stat*	*P-value*	*Lower 95%*	*Upper 95%*	*Lower 95.0%*	*Upper 95.0%*
Intercept	-1.70	0.784	-2.172	0.055	-3.451	0.044	-3.451	0.044
X Variable 1	0.03	0.005	6.086	0.000	0.020	0.043	0.020	0.043

Fig. 6.32 Example of a Chart with the Regression Equation Displayed Next to the Regression Line

$$Y = 0.03X - 1.70$$

Print this file so that it fits onto only one page.
Then, save this file as: LSAT9

6.7 How to Recognize Negative Correlations in the SUMMARY OUTPUT Table

Important note: Since Excel does not recognize negative correlations in the SUMMARY OUTPUT results, buttreats all correlations as if they were positive correlations (this was a mistake made by the programmer), you need to be careful to note that there may be a negative correlation between X and Y even if the printout says that the correlation is a positive correlation.

You will know that the correlation between X and Y is a negative correlation when these two things occur:

(1) *THE SLOPE, b, IS A NEGATIVE NUMBER. This can onlyoccur when there is a negative correlation.*
(2) *THE CHART CLEARLY SHOWS A DOWNWARD SLOPE IN THE REGRESSION LINE, which can only occur whenthe correlation between X and Y is negative.*

6.8 Printing Only Part of a Spreadsheet Instead of the Entire Spreadsheet

Objective: To print part of a spreadsheet separately instead of printing the entire spreadsheet

There will be many occasions when your spreadsheet is so large in the number of cells used for your data and charts that you only want to print part of the spreadsheet separately so that the print will not be so small that you cannot read it easily.

We will now explain how to print only part of a spreadsheet onto a separate page by using three examples of how to do that using the file, LSAT9, that you created in Sect. 6.6: (1) printing only the table and the chart on a separate page, (2) printing only the chart on a separate page, and (3) printing only the SUMMARY OUTPUT of the regression analysis on a separate page.

Note: If the file: LSAT9 is not open on your screen, you need to open it now.
If the border is around the outside of the chart, click on any white space outside the chart to deselect the chart.

Let's describe how to do these three goals with three separate objectives:

6.8.1 Printing Only the Table and the Chart on a Separate Page

Objective: To print only the table and the chart on a separate page

1. Left-click your mouse starting at the top left of the table *in cell A2*and drag the mouse *down and to the right so that all of the table and all of the chart are highlighted in light blue on your computer screen from cell A2 to cell H48* (the light blue cells are called the "selection" cells).
2. File
   Print
   Print Active Sheet (hit the down arrow on the right)
   Print selection
   Print

The resulting printout should contain only the table of the data and the chart resulting from the data.

Then, click on any empty cell in your spreadsheet to deselect the table and chart.

6.8.2 Printing Only the Chart on a Separate Page

Objective: To print only the chart on a separate page

1. Click on any "white space" *just inside the outside border of the chart in the top right corner of the chart* to create the border around all of the borders of the chart in order to "select" the chart.
2. File
   Print
   Print selected chart
   Print selected chart (again)

Print
The resulting printout should contain only the chart resulting from the data.

Important note: After each time you print a chart by itself on a separate page, you should immediately click on any white space OUTSIDE the chart to remove the border from the chart. When the border is on the borders of the chart, this tells Excel that you want to print only the chart by itself. Do this now!

6.8.3 Printing Only the SUMMARY OUTPUT of the Regression Analysis on a Separate Page

Objective: To print only the SUMMARY OUTPUT of the regression analysis on a separate page

1. Left-click your mouse at the cell just above SUMMARY OUTPUT in *cell A49* on the left of your spreadsheet and drag the mouse *down and to the right* until all of the regression output is highlighted in dark blue on your screen from A49 to I67.
2. File
   Print
   Print active sheets (hit the down arrow on the right)
   Print selection
   Print

The resulting printout should contain only the summary output of the regression analysis on a separate page.

Finally, click on any empty cell on the spreadsheet to "deselect" the regression table.

6.9 End-of-Chapter Practice Problems

1. Suppose that you wanted to study the relationship between the percent of residents in a state (age 25+) with college degrees who are public high school graduates and the percent of women who are legislators in that state. We will use hypothetical data for this problem, but if you wanted to find the actual data, you can find the educational attainment of state residents at U.S. Census Bureau (2003) and the percent of a state legislator who are women at the Center for American Women and Politics (2007). (*If you do decide to go to these Web sites*

for the actual data, be sure to use the same calendar year for both variables or your results will be incorrect!). The hypothetical data are given in Fig. 6.33:

RELATIONSHIP BETWEEN COLLEGE DEGREES AND % OF WOMEN IN STATE LEGISLATORS

X = Percent of college degrees held for public high school graduates age 25+ by state

Y = Percent of state legislators who are women

% college degrees	% female state legislators
11	9
14	18
21	35
17	31
19	36
21	28
14	18
22	40
13	18
16	12
18	15
19	12

Fig. 6.33 Worksheet Data for Chap. 6: Practice Problem #1

Create an Excel spreadsheet and enter the data *using % college degrees as the independent variable (predictor) and % female state legislators as the dependent variable (criterion).*

Important note: When you are trying to find a correlation between two variables, it is important that you place the predictor, X, ON THE LEFT COLUMN in your Excel spreadsheet, and the criterion, Y, IMMEDIATELY TO THE RIGHT OF THE X COLUMN. You should do this every time that you want to use Excel to find a correlation between two variables to check your thinking so that you do not confuse these two variables with one another.

(a) Create an *XY scatterplot* of these two sets of data such that:

- Top title: RELATIONSHIP BETWEEN COLLEGE DEGREES AND % OF FEMALE STATE LEGISLATORS

- x-axis title: Percent college degrees by state
- y-axis title: Percent female state legislators
- re-size the chart so that it is 8 columns wide and 25 rows long
- move the chart below the table

(b) Create the *least-squares regression line* for these data on the scatterplot, and add the regression equation to the chart using Excel commands.

(c) Use Excel to run the regression statistics to find the *equation for the least-squares regression line* for these data and display the results below the chart on your spreadsheet. Use number format (two decimal places) for the correlation and for all the other decimal numbers.

(d) Print just the input data and the chart so that this information fits onto one page. Then, print the regression output table on a separate page so that it fits onto that separate page.

(e) save the file as: college3

Now, answer these questions using your Excel printout:

(1) What is the y-intercept?

(2) What is the slope of the line?

(3) What is the regression equation for these data using the SUMMARY OUTPUT (use two decimal places for the y-intercept and the slope)?

(4) Use the regression equation to predict the percent female state legislators you would expect for a state that had 20 % of its residents with college degrees and who were public high school graduates.

2. Suppose that you wanted to study the relationship between DIET (measured in calories allowed per day) and WEIGHT LOSS (measured in kilograms, kg) for adult women between the ages of 30 and 40 who are overweight for their height and body structure, and who all weigh roughly the same number of kilograms before undertaking the weight loss program. You want to test your Excel skills on a random sample of these women based on their weight change over the past four months to make sure that you can do this type of research. The hypothetical data appear in Fig. 6.34:

Fig. 6.34 Worksheet Data for Chap. 6: Practice Problem #2

RELATIONSHIP BETWEEN DIET AND WEIGHT LOSS	
ADULT WOMEN AGES 30-40	
DIET (calories allowed per day)	WEIGHT LOSS (kg)
900	16.0
1050	12.0
1150	8.0
1275	6.0
1420	3.0
1530	5.5
1610	9.5
1710	2.5
1820	6.0
1875	9.0
1930	6.0
2100	3.0

Create an Excel spreadsheet and enter the data using DIET (calories allowed per day) as the independent variable (predictor) and WEIGHT LOSS (kg) as the dependent variable (criterion). Underneath the table, use Excel's = *correl* function to find the correlation between these two variables. Label the correlation and place it below the table. Then, round off the correlation to two decimal places.

(a) create an *XY scatterplot* of these two sets of data such that:

- top title: RELATIONSHIP BETWEEN DIET AND WEIGHT LOSS
- x-axis title: DIET (calories allowed per day)
- y-axis title: WEIGHT LOSS (kg)
- move the chart below the table
- re-size the chart so that it is 7 columns wide and 25 rows long

(b) Create a *least-squares regression line* for these data on the scatterplot, and add the regression equation to the chart.

(c) Use Excel to run the regression statistics to find the *equation for the least-squares regression line* for these data and display the results below the chart on your spreadsheet. Use number format (two decimal places) for the correlation, r, and three decimal places for all of the other decimal figures, including the coefficients.

(d) Print the input data and the chart so that this information fits onto one page.

(e) Then, print out the regression output table so that this information fits onto a separate page.

(f) Save this file as: DIET3

Answer the following questions using your Excel printout:

1. What is the correlation between DIET and WEIGHT LOSS?
2. What is the y-intercept?
3. What is the slope of the line?
4. What is the regression equation for these data?
5. Use that regression equation to predict the WEIGHT LOSS you would expect for a woman who was practicing a DIET of 1500 calories allowed a day. Show your work on a separate sheet of paper.

(Note that this correlation is not the multiple correlation as the Excel table indicates, but is merely the correlation r instead.)

Important note: Since Excel does not recognize negative correlations in the SUMMARY OUTPUT but treats all correlations as if they were positive correlations, you need to be careful to note that there is a negative correlation between DIET and WEIGHT LOSS.

You know this for two reasons:

(1) *The slope, b, is a negative −0.64 which can only occur when there is a negative correlation.*
(2) *The chart clearly shows a downward slope in the regression line, which can only happen when the correlation is negative.*

Therefore, the correlation between DIET and WEIGHT LOSS is not +0.64, but −0.64 for this problem.
This is a negative correlation!

3. Suppose that you wanted to study the relationship between the population density of a state (e.g. the average number of people per square mile) and the number of traffic fatalities in that state. Would you expect this correlation to be positive or negative? We will be using hypothetical data for this problem, but you can find the actual data for population density at the Web site for the U.S. Census Bureau (2005) and the actual data for traffic fatalities in a state at a different Web site for the U.S. Census Bureau (2008). (*If you do decide to go to these Web sites for the actual data, be sure to use the same calendar year for both variables or your results will be incorrect*!). The hypothetical data appear in Fig. 6.35:

RELATIONSHIP BETWEEN STATE POPULATION DENSITY AND TRAFFIC FATALITIES

X = State population density per square mile

Y = Motor vehicle fatalities within 30 days of the accident

Population density	Motor vehicle fatalities
17	65
25	123
80	612
124	136
724	3400
653	1800
541	613
376	512
412	316
114	363
52	163
91	25

Fig. 6.35 Worksheet Data for Chap. 6: Practice Problem #3

Create an Excel spreadsheet and enter the data using the population density of a state as the independent variable (predictor) and the motor vehicle fatalities as the dependent variable (criterion).

(a) Use Excel's = correl function to find the correlation between these two variables underneath the table you create, label it, and round if off to two decimal places.

(b) create an *XY scatterplot* of these two sets of data such that:

- top title: RELATIONSHIP BETWEEN POPULATION DENSITY AND TRAFFIC FATALITIES
- x-axis title: STATE POPULATION DENSITY
- y-axis title: TRAFFIC FATALITIES
- move the chart below the table
- re-size the chart so that it is 8 columns wide and 25 rows long

(c) Create the *least-squares regression line* for these data on the scatterplot, and add the regression equation to the chart.

(d) Use Excel to run the regression statistics to find the *equation for the least-squares regression line* for these data and display the results below the chart on your spreadsheet. Use number format (two decimal places) for the correlation and the two coefficients, and use two decimal places for all other decimal figures.

(e) Print just the input data and the chart so that this information fits onto one page. Then, print the regression output table on a separate page so that it fits onto that separate page.

(f) save the file as: Vehicle13

Answer the following questions using your Excel printout:

1. What is the correlation between the state population density and the traffic fatalities?
2. What is the y-intercept?
3. What is the slope of the line?
4. What is the regression equation using the SUMMARY OUTPUT?
5. Use the regression equation to predict the traffic fatalities you would expect for a state that had a population density of 670 people per square mile. Show your work on a separate sheet of paper.

References

Black, K. Business Statistics: For Contemporary Decision Making (6th ed.). Hoboken, NJ: John Wiley & Sons, Inc., 2010.

Center for American Women and Politics. Retrieved September 2, 2011 from http://www.cawp.rutgers.edu/Facts/Officeholders/stleg.pdf, 2007.

Johnson, J.B. and Reynolds, H.T. Political Science Research Methods (5th ed.). Washington D.C.: CQ Press, 2005.

King, G., Keohane, R.O., and Verba, S. Designing Social Inquiry: Scientific Inference in Qualitative Research. Princeton, NJ: Princeton University Press, 1994.

Levine, D.M.. Stephan, D.F., Krehbiel, T.C., and Berenson, M.L. Statistics for Managers Using Microsoft Excel (6th ed.). Boston, MA: Prentice Hall/Pearson, 2011.

Shively, W.P. The Craft of Political Research (7th ed.). Upper Saddle River, NJ: Prentice Hall/ Pearson, 2009.

Steinberg, W.J. Statistics Alive! Thousand Oaks, CA: Sage Publications, 2008.

Zikmund, W.G. and Babin, B.J. Exploring Marketing Research (10th ed.). Mason, OH: South-Western Cengage Learning, 2010.

U.S. Census Bureau. Retrieved September 3, 2011 from http://www.census.gov/compendia/smadb/TableA-22.pdf, 2003.

U.S. Census Bureau. Retrieved September 2, 2011 from http://www.census.gov/compendia/smadb/TableA-01.pdf, 2005.

U.S. Census Bureau. Retrieved September 4, 2011 from http://www.census.gov/compendia/statab/cats/transportaion/motor_vehicle_accidents_and_fatalities.html, Table 1103, 2008.

Chapter 7
Multiple Correlation and Multiple Regression

There are many times in social science research when you want to predict a criterion, Y, but you want to find out if you can develop a better prediction model by using *several predictors* in combination (e.g. X_1, X_2, X_3, etc.) instead of a single predictor, X.

The resulting statistical procedure is called "multiple correlation" because it uses two or more predictors in combination to predict Y, instead of a single predictor, X. Each predictor is "weighted" differently based on its separate correlation with Y and its correlation with the other predictors. The job of multiple correlation is to produce a regression equation that will weight each predictor differently and in such a way that the combination of predictors does a better job of predicting Y than any single predictor by itself. We will call the multiple correlation: R_{xy}.

Important note: Note that R_{xy} ranges from zero to +1.0. It is never negative!

You will recall (see Sect. 6.5.3) that the regression equation that predicts Y when only one predictor, X, is used is:

$$Y = a + bX \tag{7.1}$$

7.1 Multiple Regression Equation

The multiple regression equation follows a similar format and is:

$$Y = a + b_1X_1 + b_2X_2 + b_3X_3 + \text{ etc. depending on the number of predictors used} \tag{7.2}$$

The "weight" given to each predictor in the equation is represented by the letter "b" with a subscript to correspond to the same subscript on the predictors.

© Springer International Publishing Switzerland 2016

T.J. Quirk, *Excel 2016 for Social Science Statistics*, Excel for Statistics,
DOI 10.1007/978-3-319-39711-5_7

Important note: In order to do multiple regression, you need to have installed the "Data Analysis TookPak" that was described in Chap. 6 (see Sect. 6.5.1). If you did not install this, you need to do so now.

Let's try a practice problem.

Suppose that you have been asked to analyze some data from the SAT Reasoning Test (formerly called the Scholastic Aptitude Test) which is a standardized test for college admissions in the U.S. This test is intended to measure a student's readiness for academic work in college, and about 1.4 million high school students take this test every year. There are three subtest scores generated from this test: Critical Reading, Writing, and Mathematics, and each of these subtests has a score range between 200 and 800 with an average score of about 500.

Suppose that a nearby selective liberal arts college in the northeast of the U.S. that is near to you wants to determine the relationship between SAT Reading scores, SAT Writing scores, and SAT Math scores in their ability to predict freshman grade-point average (FROSH GPA) at the end of freshman year at this college, and that this college has asked you to determine this relationship.

You have decided to use the three subtest scores as the predictors, X_1, X_2, and X_3 and the freshman grade-point average (FROSH GPA) as the criterion, Y. To test your Excel skills, you have selected 11 students randomly from last year's freshmen class, and have recorded their scores on these variables.

Let's use the following notation:

Y	FROSH GPA
X_1	READING SCORE
X_2	WRITING SCORE
X_3	MATH SCORE

Suppose, further, that you have collected the following hypothetical data summarizing these scores (see Fig. 7.1):

SAT REASONING TEST

Is there a relationahip between SAT scores and Freshman GPA at a local college?

FROSH GPA	READING SCORE	WRITING SCORE	MATH SCORE
2.55	250	230	220
3.05	610	240	440
3.55	620	540	530
2.05	420	420	260
2.45	320	520	320
2.95	630	620	620
3.15	650	540	530
3.45	520	580	560
3.30	420	490	630
2.75	330	220	610
3.65	440	570	660

Fig. 7.1 SAT versus FROSH GPA (Practical Example)

Create an Excel spreadsheet for these data using the following cell reference:

A2: SAT REASONING TEST
A4: Is there a relationship between SAT scores and Freshman GPA at a local college?
A6: FROSH GPA
A7: 2.55
B6: READING SCORE
C6: WRITING SCORE
D6: MATH SCORE
D17: 660

Next, change the column width to match the above table, and change all GPA figures to number format (two decimal places).

Now, fill in the additional data in the chart such that:

A17: 3.65
B17: 440
C17: 570

Then, center all numbers in your table

Important note: Be sure to double-check all of your numbers in your table to be sure that they are correct, or your spreadsheets will be incorrect.

Save this file as: GPA15

Before we do the multiple regression analysis, we need to try to make one important point very clear:

Important: When we used one predictor, X, to predict one criterion, Y, we said that you need to make sure that the X variable is ON THE LEFT in your table, and the Y variable is ON THE RIGHT in your table so that you don't get these variables mixed up (see Sect. 6.3).

However, in multiple regression, you need to follow this rule which is exactly the opposite:

When you use several predictors in multiple regression, it is essential that the criterion you are trying to predict, Y, be ON THE FAR LEFT, and all of the predictors are TO THE RIGHT of the criterion, Y, in your table so that you know which variable is the criterion, Y, and which variables are the predictors. If you make this a habit, you will save yourself a lot of grief.

Notice in the table above, that the criterion Y (FROSH GPA) is on the far left of the table, and the three predictors (READING SCORE, WRITING SCORE, and MATH SCORE) are to the right of the criterion variable. If you follow this rule, you will be less likely to make a mistake in this type of analysis.

7.2 Finding the Multiple Correlation and the Multiple Regression Equation

Objective: To find the multiple correlation and multiple regression equation using Excel.

You do this by the following commands:
Data
Click on: Data Analysis (far right top of screen)
Regression (scroll down to this in the box; see Fig. 7.2)

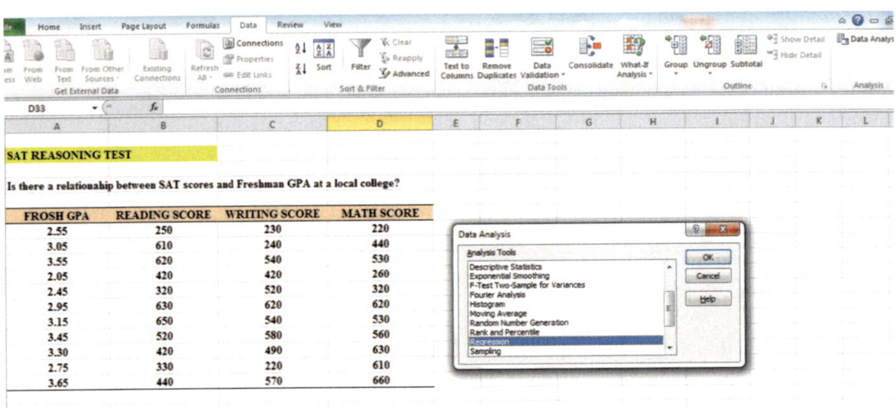

Fig. 7.2 Dialogue Box for Regression Function

OK

Input Y Range: A6:A17
Input X Range: B6:D17

Click on the Labels box to *add a check mark* to it (because you have included the column labels in row 6)

Output Range (click on the button to its left, and enter): A20 (see Fig. 7.3)

Important note: Excel automatically assigns a dollar sign $ in front of each column letter and each row number so that you can keep these ranges of data constant for the regression analysis.

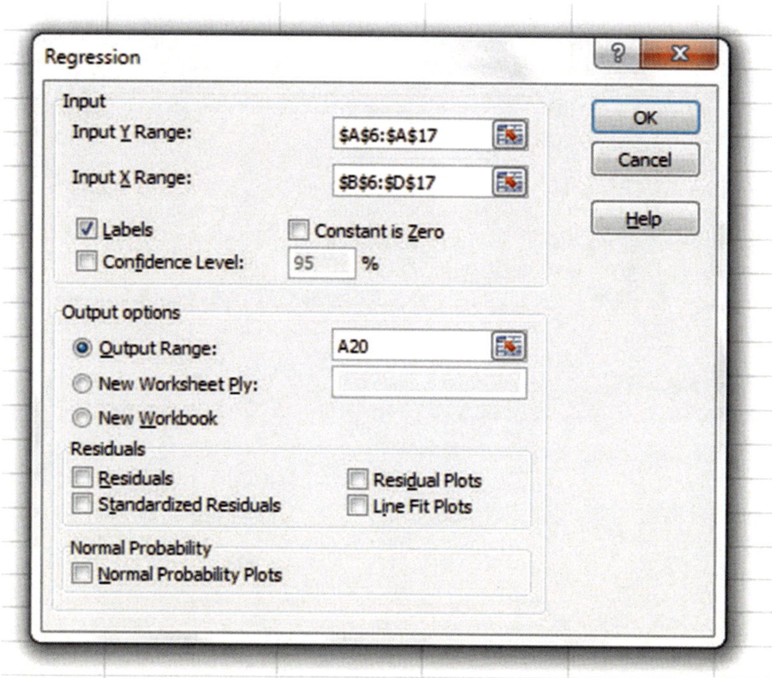

Fig. 7.3 Dialogue Box for SAT vs. FROSH GPA Data

OK (see Fig. 7.4 to see the resulting SUMMARY OUTPUT)
Next, format cell B23 in number format (two decimal places)

	A	B	C	D	E	F	G	H	I
19									
20	SUMMARY OUTPUT								
21									
22	*Regression Statistics*								
23	Multiple R	0.797651156							
24	R Square	0.636247366							
25	Adjusted R Square	0.48035338							
26	Standard Error	0.361446932							
27	Observations	11							
28									
29	ANOVA								
30		*df*	*SS*	*MS*	*F*	*Significance F*			
31	Regression	3	1.599583719	0.533194573	4.081282	0.057174747			
32	Residual	7	0.91450719	0.130643884					
33	Total	10	2.514090909						
34									
35		*Coefficients*	*Standard Error*	*t Stat*	*P-value*	*Lower 95%*	*Upper 95%*	*Lower 95.0%*	*Upper 95.0%*
36	Intercept	1.53627108	0.468442063	3.279532734	0.013496	0.428581617	2.643960543	0.428581617	2.643960543
37	READING SCORE	0.000642945	0.000963026	0.667629207	0.525762	-0.001634251	0.00292014	-0.001634251	0.00292014
38	WRITING SCORE	0.000264354	0.000889915	0.297055329	0.775046	-0.00183996	0.002368667	-0.00183996	0.002368667
39	MATH SCORE	0.00210733	0.000848684	2.4830572	0.042022	0.000100512	0.004114149	0.000100512	0.004114149
40									

Fig. 7.4 Regression SUMMARY OUTPUT of SAT vs. FROSH GPA Data

Next, format the following four cells in Number format (four decimal places):

B36
B37
B38
B39

Change all other decimal figures to two decimal places, and center all figures within their cells.

Note that both the input Y Range and the Input X Range above both include the label at the top of the columns.

Save the file as: GPA16

Now, print the file so that it fits onto one page by changing the scale to *60 % size*. The resulting regression analysis is given in Fig. 7.5.

SAT REASONING TEST

Is there a relationship between SAT scores and Freshman GPA at a local college?

FROSH GPA	READING SCORE	WRITING SCORE	MATH SCORE
2.55	250	230	220
3.05	610	240	440
3.55	620	540	530
2.05	420	420	260
2.45	320	520	320
2.95	630	620	620
3.15	650	540	530
3.45	520	580	560
3.30	420	490	630
2.75	330	220	610
3.65	440	570	660

SUMMARY OUTPUT

Regression Statistics	
Multiple R	0.80
R Square	0.64
Adjusted R Square	0.48
Standard Error	0.36
Observations	11

ANOVA

	df	*SS*	*MS*	*F*	*Significance F*
Regression	3	1.60	0.53	4.08	0.06
Residual	7	0.91	0.13		
Total	10	2.51			

	Coefficients	*Standard Error*	*t Stat*	*P-value*	*Lower 95%*	*Upper 95%*	*Lower 95.0%*	*Upper 95.0%*
Intercept	1.5363	0.47	3.28	0.01	0.43	2.64	0.43	2.64
READING SCORE	0.0006	0.00	0.67	0.53	0.00	0.00	0.00	0.00
WRITING SCORE	0.0003	0.00	0.30	0.78	0.00	0.00	0.00	0.00
MATH SCORE	0.0021	0.00	2.48	0.04	0.00	0.00	0.00	0.00

Fig. 7.5 Final Spreadsheet for SAT vs. FROSH GPA Regression Analysis

Once you have the SUMMARY OUTPUT, you can determine the multiple correlation and the regression equation that is the best-fit line through the data points using READING SCORE, WRITING SCORE, AND MATH SCORE as the three predictors, and FROSH GPA as the criterion.

Note on the SUMMARY OUTPUT where it says: "Multiple R." This term is correct since this is the term Excel uses for the multiple correlation, which is +0.80. This means, that from these data, that the combination of READING SCORES, WRITING SCORES, AND MATH SCORES together form a very strong positive relationship in predicting FROSH GPA.

To find the regression equation, *notice the coefficients at the bottom of the SUMMARY OUTPUT*:

Intercept : *a* (*this is the y-intercept*)	*1.5363*
READING SCORE: b_1	*0.0006*
WRITING SCORE: b_2	*0.0003*
MATH SCORE: b_3	0.0021

Since the general form of the multiple regression equation is:

$$Y = a + b_1X_1 + b_2X_2 + b_3X_3 \tag{7.2}$$

we can now write the multiple regression equation for these data:

$$Y = 1.5363 + 0.0006X_1 + 0.0003X_2 + 0.0021X_3$$

7.3 Using the Regression Equation to Predict FROSH GPA

Objective: To find the predicted FROSH GPA using an SAT Reading Score of 600, an SAT Writing Score of 500, and an SAT Math Score of 550

Plugging these three numbers into our regression equation gives us:

$$\mathrm{Y} = 1.5363 + 0.0006(600) + 0.0003(500) + 0.0021(550)$$
$$\mathrm{Y} = 1.5363 + 0.36 + 0.15 + 1.155$$
$$\mathrm{Y} = 3.20 \text{ (since GPA scores are typically measured in two decimals)}$$

If you want to learn more about the theory behind multiple regression, see Keller (2009) and Shively (2009).

7.4 Using Excel to Create a Correlation Matrix in Multiple Regression

The final step in multiple regression is to find the correlation between all of the variables that appear in the regression equation.

In our example, this means that we need to find the correlation between each of the six pairs of variables:

To do this, we need to use Excel to create a "correlation matrix." This matrix summarizes the correlations between all of the variables in the problem.

Objective: To use Excel to create a correlation matrix between the four variables in this example.

To use Excel to do this, use these steps:
Data (top of screen under "Home" at the top left of screen)
Data Analysis
Correlation (scroll *up* to highlight this formula; see Fig. 7.6)

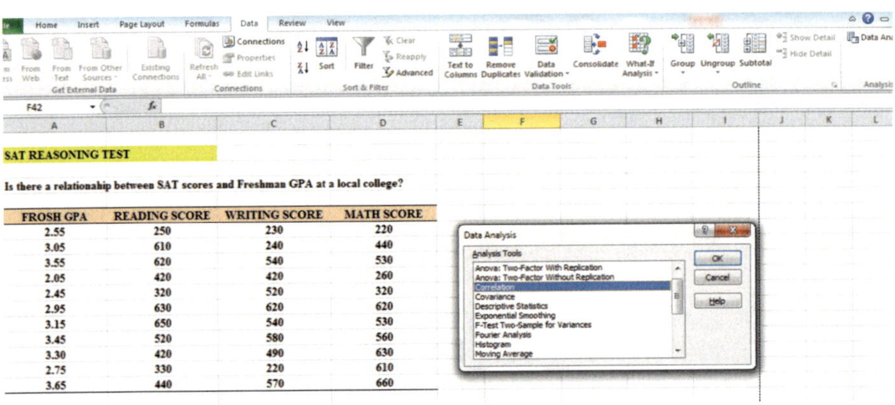

Fig. 7.6 Dialogue Box for SAT vs. FROSH GPA Correlations

OK

Input range: A6:D17

(Note that this input range includes the labels at the top of the FOUR variables (FROSH GPA, READING SCORE, WRITING SCORE, MATH SCORE) as well as all of the figures in the original data set.)
Grouped by: Columns
Put a check in the box for: Labels in the First Row (since you included the labels at the top of the columns in your input range of data above)
Output range (click on the button to its left, and enter): A42 (see Fig. 7.7)

OK

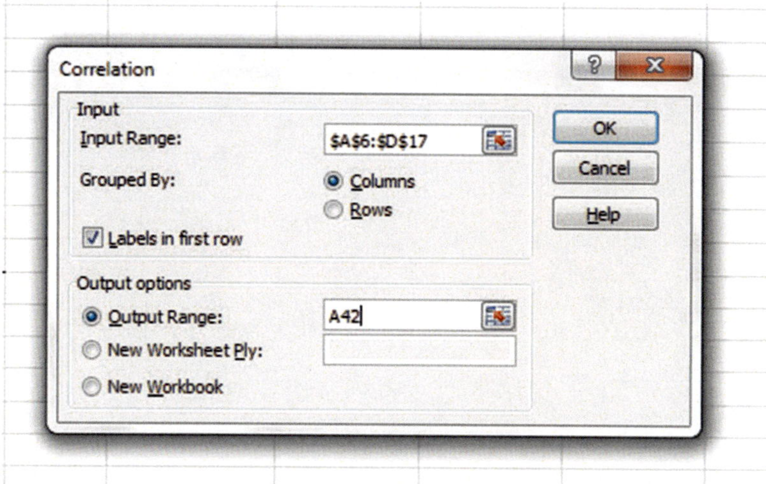

Fig. 7.7 Dialogue Box for Input/Output Range for Correlation Matrix

41					
42		*FROSH GPA*	*READING SCORE*	*WRITING SCORE*	*MATH SCORE*
43	FROSH GPA	1			
44	READING SCORE	0.510369686	1		
45	WRITING SCORE	0.446857676	0.468105152	1	
46	MATH SCORE	0.772523347	0.444074496	0.429202393	1
47					
48					

Fig. 7.8 Resulting Correlation Matrix for SAT Scores vs. FROSH GPA Data

The resulting correlation matrix appears in A42:E46 (see Fig. 7.8).

Next, format all of the numbers in the correlation matrix that are in decimals to two decimals places. And, also, make column E wider so that the MATH SCORE label fits inside cell E42.

Save this Excel file as: GPA14

The final spreadsheet for these scores appears in Fig. 7.9.

SAT REASONING TEST

Is there a relationahip between SAT scores and Freshman GPA at a local college?

FROSH GPA	READING SCORE	WRITING SCORE	MATH SCORE
2.55	250	230	220
3.05	610	240	440
3.55	620	540	530
2.05	420	420	260
2.45	320	520	320
2.95	630	620	620
3.15	650	540	530
3.45	520	580	560
3.30	420	490	630
2.75	330	220	610
3.65	440	570	660

SUMMARY OUTPUT

Regression Statistics	
Multiple R	0.80
R Square	0.64
Adjusted R Square	0.48
Standard Error	0.36
Observations	11

ANOVA

	df	*SS*	*MS*	*F*	*Significance F*
Regression	3	1.60	0.53	4.08	0.06
Residual	7	0.91	0.13		
Total	10	2.51			

	Coefficients	*Standard Error*	*t Stat*	*P-value*	*Lower 95%*	*Upper 95%*	*Lower 95.0%*	*Upper 95.0%*
Intercept	1.5363	0.47	3.28	0.01	0.43	2.64	0.43	2.64
READING SCORE	0.0006	0.00	0.67	0.53	0.00	0.00	0.00	0.00
WRITING SCORE	0.0003	0.00	0.30	0.78	0.00	0.00	0.00	0.00
MATH SCORE	0.0021	0.00	2.48	0.04	0.00	0.00	0.00	0.00

	FROSH GPA	*READING SCORE*	*WRITING SCORE*	*MATH SCORE*
FROSH GPA	1			
READING SCORE	0.51	1		
WRITING SCORE	0.45	0.47	1	
MATH SCORE	0.77	0.44	0.43	1

Fig. 7.9 Final Spreadsheet for SAT Scores vs. FROSH GPA Regression and the Correlation Matrix

Note that the number "1" along the diagonal of the correlation matrix means that the correlation of each variable with itself is a perfect, positive correlation of 1.0.

Correlation coefficients are always expressed in just two decimal places.

You are now ready so read the correlation between the six pairs of variables:

The correlation between READING SCORE and FROSH GPA is: *+.51*
The correlation between WRITING SCORE and FROSH GPA is: *+.45*
The correlation between MATH SCORE and FROSH GPA is: *+.77*
The correlation between WRITING SCORE and READING SCORE is: *+.47*
The correlation between MATH SCORE and READING SCORE is: *+.44*
The correlation between MATH SCORE and WRITING SCORE is: *+.43*

This means that the best predictor of FROSH GPA is the MATH SCORE with a correlation of +.77. Adding the other two predictor variables, READING SCORE and WRITING SCORE, improved the prediction by only 0.03 to 0.80, and was, therefore, only slightly better in prediction. MATH SCORES are an excellent predictor of FROSH GPA all by themselves.

If you want to learn more about the correlation matrix, see Levine et al. (2011).

7.5 End-of-Chapter Practice Problems

1. Suppose that you wanted to study the relationship between the population density per square mile in a state, the percent of the state's population age 18–24, and motor vehicle fatalities (per 100,000 population) in the state. U.S. insurance companies charge a high premium price for auto insurance for drivers age 18–24 because statistically these drivers are more likely to be involved in a vehicle accident than older drivers. Essentially, insurance companies are hypothesizing that: (1) there is a negative correlation between population density and motor vehicle fatalities because less populated areas are more likely to be farther from medical assistance, and (2) there is a positive correlation between the percent of the population age 18–24 and motor vehicle fatalities because the more younger drivers a state has statistically, the more likely these drivers are to be involved in motor vehicle accidents than older drivers. We will be using hypothetical data for this problem, but if you want to find the actual data for the percent of a state's population age 18–24, you can check out the website for the U.S. Census Bureau (2004). Data for the population density for a state can be found at the U.S. Census Bureau Web site (2005).

Important note: *If you do decide to find the actual data be sure to use the same calendar year for all three variables* (e.g. *2015*) *or your results will be incorrect.*

The hypothetical data are given in Fig. 7.10:

MOTOR VEHICLE FATALITIES IN TERMS OF POPULATION DENSITY AND AGE OF RESIDENTS

Question: What is the relationship between the population density of a state, the percent of that state's population ages 18-24, and motor vehicle fatalities in that state?

	Y	X1	X2
State	Motor vehicle fatalities (per 100,000 population)	Population density (per square mile)	Percent of population (ages 18 - 24)
1	18	80	5
2	6	120	3
3	5	200	4
4	7	220	4
5	8	175	6
6	7	195	6
7	12	50	8
8	14	70	9
9	8	185	4
10	6	190	6

Fig. 7.10 Worksheet Data for Chap. 7: Practice Problem #1

(a) create an Excel spreadsheet using the motor vehicle fatalities figures as the criterion and the population density and the percent of the population age 18–24 figures as the predictors.

(b) Use Excel's *multiple regression* function to find the relationship between these three variables and place the SUMMARY OUTPUT below the table.

(c) Use number format (two decimal places) for the multiple correlation on the Summary Output, and use this same number format for the coefficients in the summary output.

(d) Save the file as: Vehicle23

(e) Print the table and regression results below the table so that they fit onto one page.

Answer the following questions using your Excel printout:

1. What is multiple correlation R_{xy}?
2. What is the y-intercept a?
3. What is the coefficient for population density b_1?
4. What is the coefficient for percent of population age 18–24 b_2?
5. What is the multiple regression equation?
6. Predict the motor vehicle fatalities you would expect for a state with a population density of 220 and which had 8 % of its population age 18–24.

(f) Now, go back to your Excel file and create a correlation matrix for these three variables, and place it underneath the SUMMARY OUTPUT on your spreadsheet.
(g) Save this file as: Vehicle23
(h) Now, print out *just this correlation matrix* on a separate sheet of paper. Answer the following questions using your Excel printout. Be sure to include the plus or minus sign for each correlation:

7. What is the correlation between population density and vehicle fatalities?
8. What is the correlation between percent of population age 18–24 and vehicle fatalities?
9. What is the correlation between population density and percent of population age 18–24?
10. Discuss which of the two predictors is the better predictor of vehicle fatalities:
11. Explain in words how much better the two predictor variables combined predict vehicle fatalities than the best single predictor.

2. The Graduate Record Examinations (GRE) are a standardized test that is an admissions requirement for many U.S. graduate schools. The test is intended to measure general academic preparedness, regardless of specialization field. The General GRE test produces three subtest scores: (1) GRE Verbal Reasoning (scale 130–170), (2) GRE Quantitative Reasoning (scale 130–170), and (3) Analytical Writing (scale 0–6).

   Suppose that you have been asked by the Chair of the Psychology Department at a selective graduate school to see how well the GRE predicts GPA at the end of the first year of graduate study in Psychology. This Chair has asked you to use the three subtest scores of the GRE as predictors, and, in addition, to use the GRE Psychology Subject Test score (range 200–990) as an additional predictor of this GPA. The Chair would like your recommendation as to whether or not the Psychology Test should become an admissions requirement in addition to the GRE for admission to the graduate program in psychology. About 7,000 prospective students take the GRE Psychology Test each year.

   You have decided to use a multiple correlation and multiple regression analysis, and to test your Excel skills, you have collected the data of a random sample of 12 Psychology students who have just finished their first year of graduate study at this university. These hypothetical data appear in Fig. 7.11:

GRADUATE RECORD EXAMINATIONS				
How well does the GRE and the GRE subject area test in Psychology predict GPA at the end of the first year of a Masters' program in Psychology?				
FIRST-YEAR GPA	GRE VERBAL	GRE QUANTITATIVE	GRE WRITING	GRE PSYCHOLOGY
3.25	160	161	5	650
3.42	156	158	4	600
2.85	156	157	2	500
2.65	154	153	1	510
3.65	166	166	6	630
3.16	159	160	3	550
3.56	166	163	4	610
2.35	155	154	2	430
2.86	153	154	3	450
2.95	158	157	4	550
3.15	158	159	4	580
3.45	160	160	5	620

Fig. 7.11 Worksheet Data for Chap. 7: Practice Problem #2

(a) create an Excel spreadsheet using FIRST-YEAR GPA as the criterion (Y), and GRE VERBAL (X_1), GRE QUANTITATIVE (X_2), GRE WRITING (X_3), and GRE PSYCHOLOGY (X_4) as the predictors.

(b) Use Excel's *multiple regression* function to find the relationship between these five variables and place it below the table.

(c) Use number format (two decimal places for the multiple correlation on the SUMMARY OUTPUT, and use four decimal places for the coefficients in the SUMMARY OUTPUT.

(d) Print the table and regression results below the table so that they fit onto one page.

(e) Save this file as: GRE36

Answer the following questions using your Excel printout:

1. What is the multiple correlation R_{xy}?
2. What is the y-intercept a?
3. What is the coefficient for GRE VERBAL b_1?
4. What is the coefficient for GRE QUANTITATIVE b_2?
5. What is the coefficient for GRE WRITING b_3?
6. What is the coefficient for GRE PSYCHOLOGY b_4?
7. What is the multiple regression equation?
8. Predict the FIRST-YEAR GPA you would expect for a GRE VERBAL score of 160, a GRE QUANTITATIVE score of 150, a GRE WRITING score of 3, and a GRE PSYCHOLOGY score of 610.

(f) Now, go back to your Excel file and create a *correlation matrix* for these five variables, and place it underneath the SUMMARY OUTPUT.

(g) Re-save this file as: GRE36

(h) Now, print out *just this correlation matrix* on a separate sheet of paper. Answers the following questions using your Excel printout. Be sure to include the plus or minus sign for each correlation:

9. What is the correlation between GRE VERBAL and FIRST-YEAR GPA?
10. What is the correlation between GRE QUANTITATIVE and FIRST-YEAR GPA?
11. What is the correlation between GRE WRITING and FIRST-YEAR GPA?
12. What is the correlation between GRE PSYCHOLOGY and FIRST-YEAR GPA?
13. What is the correlation between GRE WRITING and GRE VERBAL?
14. What is the correlation between GRE VERBAL and GRE PSYCHOLGY?
15. Discuss which of the four predictors is the best predictor of FIRST-YEAR GPA.
16. Explain in words how much better the four predictor variables together predict FIRST-YEAR GPA than the best single predictor by itself.

3. Suppose that you are the marketing manager for 7Eleven Stores in Missouri and that you want to see if a proposed store location would generate sufficient yearly sales volume to support the idea of building a new store at that location. You have checked the data available at your company to generate the following table for a random sample of 20 7Eleven stores in Missouri based on last year's data to create the hypothetical data given in Fig. 7.12.

	Y	X_1	X_2	X_3
Store ID	Annual Sales ($000)	Average Daily Traffic	Population (2-mile radius)	Average Income in Area
1	1,121	61,655	17,880	$28,991
2	766	35,236	13,742	$14,731
3	595	35,403	19,741	$8,114
4	899	52,832	23,246	$15,324
5	915	40,809	24,485	$11,438
6	782	40,820	20,410	$11,730
7	833	49,147	28,997	$10,589
8	571	24,953	9,981	$10,706
9	692	40,828	8,982	$23,591
10	1,005	39,195	18,814	$15,703
11	589	34,574	16,941	$9,015
12	671	26,639	13,319	$10,065
13	903	55,083	21,482	$17,365
14	703	37,892	26,524	$7,532
15	556	24,019	14,412	$6,950
16	657	27,791	13,896	$9,855
17	1,209	53,438	22,444	$21,589
18	997	54,835	18,096	$22,659
19	844	32,919	16,458	$12,660
20	883	29,139	16,609	$11,618

Fig. 7.12 Worksheet Data for Chap. 7: Practice Problem #3

(a) create an Excel spreadsheet using the annual sales figures as the criterion and the average daily traffic, population, and income figures as the predictors.
(b) Use Excel's *multiple regression* function to find the relationship between these four variables and place the SUMMARY OUTPUT below the table.
(c) Use number format (two decimal places) for the multiple correlation on the Summary Output, and use this same number format for the coefficients in the summary output.
(d) Save the file as: multiple2
(e) Print the table and regression results below the table so that they fit onto one page.
Answer the following questions using your Excel printout:

1. What is multiple correlation R_{xy}?
2. What is the y-intercept a?
3. What is the coefficient for Average Daily Traffic b_1?
4. What is the coefficient for Population b_2?
5. What is the coefficient for Average Income b_3?
6. What is the multiple regression equation?
7. Predict the annual sales you would expect for Average Daily Traffic of 42,000, a population of 23,000, and income of $22,000.

(f) Now, go back to your Excel file and create a correlation matrix for these four variables, and place it underneath the SUMMARY OUTPUT on your spreadsheet.
(g) Save this file as: multiple3
(h) Now, print out *just this correlation matrix* on a separate sheet of paper. Answer the following questions using your Excel printout. Be sure to include the plus or minus sign for each correlation:

8. What is the correlation between traffic and sales?
9. What is the correlation between population and sales?
10. What is the correlation between income and sales?
11. What is the correlation between traffic and population?
12. What is the correlation between population and income?
13. Discuss which of the three predictors is the best predictor of annual sales:
14. Explain in words how much better the three predictor variables combined predict annual sales than the best single predictor by itself.

References

Keller, G. Statistics for Management and Economics (8th ed.). Mason, OH: South-Western Cengage Learning, 2009.

Levine, D.M., Stephan, D.F., Krehbiel, T.C., and Berenson, M.L. Statistics for Managers using Microsoft Excel (6th ed.). Boston, MA: Prentice Hall/Pearson, 2011.

Shively, W.P. The Craft of Political Research (7th ed.). Upper Saddle River, NJ: Prentice Hall/Pearson, 2009.

U.S. Census Bureau. Retrieved September 6, 2011 from http://www.census/gov/compendia/smadb/TableA-04.pdf, 2004.

U.S. Census Bureau. Retrieved September 6, 2011 from http://www.census.gov/compendia/smadb/TableA-01.pdf, 2005.

Chapter 8
One-Way Analysis of Variance (ANOVA)

So far in this 2016 Excel Guide, you have learned how to use a one-group t-test to compare the sample mean to the population mean, and a two-group t-test to test for the difference between two sample means. *But what should you do when you have more than two groups and you want to determine if there is a significant difference between the means of these groups?*

The answer to this question is: *Analysis of Variance (ANOVA).*

The ANOVA test allows you to test for the difference between the means when you have *three or more groups* in your research study.

Important note: In order to do One-way Analysis of Variance, you need to have installed the "Data Analysis Toolpak" that was described in Chap. 6 (see Sect. 6.5.1). If you did not install this, you need to do that now.

Let's suppose that you are a research assistant for a political science professor who wants to study the attitudes of registered voters in Missouri toward military spending and a variety of other important issues in the U.S. This professor has developed a phone survey about registered voter attitudes toward the U.S. federal budget in a number of categories. She has asked you to analyze the data from the results of this survey, but you want to test your skills on a small sample of registered voters before you tackle the large data base of the results of the phone survey. You have taken a random sample of the registered voter results for the one question (Item #12) dealing with military spending, and the hypothetical data appear in Fig. 8.1: Note that each group of registered voters can be of a different number of registered voters in order for ANOVA to be used on the data. Statisticians delight in

© Springer International Publishing Switzerland 2016

T.J. Quirk, *Excel 2016 for Social Science Statistics*, Excel for Statistics,
DOI 10.1007/978-3-319-39711-5_8

US. MILITARY SPENDING

Item #12: U.S. Military spending should:

1	2	3	4	5	6	7	8	9
be decreased significantly				remain about the same				be increased significantly

Republicans	Democrats	Independents
5	3	3
8	6	6
7	4	5
4	5	4
6	7	6
9	4	5
4	5	4
6	3	6
5	2	3
7	5	4
8	4	
6	5	
7		
5		

Fig. 8.1 Worksheet Data for Attitude toward U.S. Military Spending (Practical Example)

this fact by referring to this characteristic by stating that: "ANOVA is a very robust test." (Statisticians love that term!)

Create an Excel spreadsheet for these data in this way:

A3: U.S. MILITARY SPENDING
A5: Item #12:
B5: U.S. Military spending should:
A7: 1
B7: 2

C7: 3
D7: 4
E7: 5
F7: 6
G7: 7
H7: 8
I7: 9
A8: be decreased
E8: remain about
I8: be increased
A9: significantly
E9: the same
I9: significantly
C12: Republicans
D12: Democrats
E12: Independents
C13: 5
C26: 5

Enter the other information into your spreadsheet table. When you have finished entering these data, the last cell on the left should have 5 in cell C26, and the last cell on the right should have 4 in cell E22. Center the numbers in each of the columns. Use number format (zero decimals) for all numbers.

Important note: Be sure to double-check all of your figures in the table to make sure that they are exactly correct or you will not be able to obtain the correct answer for this problem!

Save this file as: Military11

8.1 Using Excel to Perform a One-Way Analysis of Variance (ANOVA)

Objective: To use Excel to perform a one-way ANOVA test.

You are now ready to perform an ANOVA test on these data using the following steps:

Data (at top of screen)
Data Analysis (far right at top of screen)
Anova: Single Factor (*scroll up to this formula and highlight it*; see Fig. 8.2)

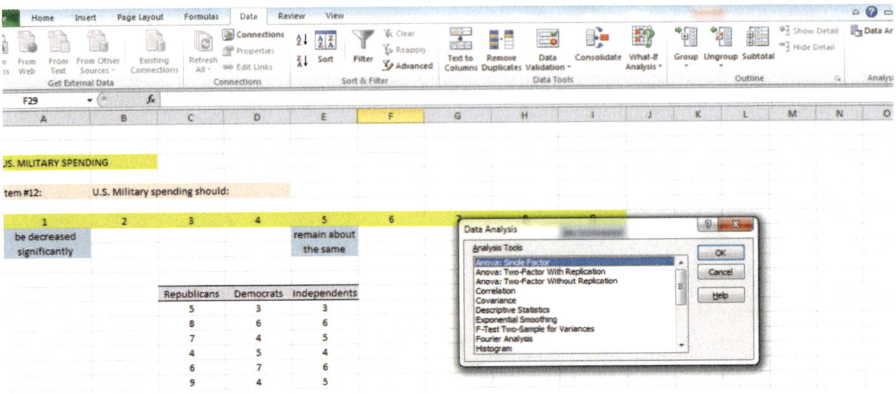

Fig. 8.2 Dialog Box for Data Analysis: Anova Single Factor

OK

Input range: C12: E26 (note that you have included in this range the column titles that are in row 12)

Important note: *Whenever the data set has a different sample size in the groups being compared, the INPUT RANGE that you define must start at the column title of the first group on the left and go to the last column on the right and go down to the lowest row that has a figure in it in the entire data matrix so that the INPUT RANGE has the "shape" of a rectangle when you highlight it. You most do this or your results will be incorrect!*

Grouped by: Columns

Put a check mark in: Labels in First Row

Output range (click on the button to its left): A28 (see Fig. 8.3)

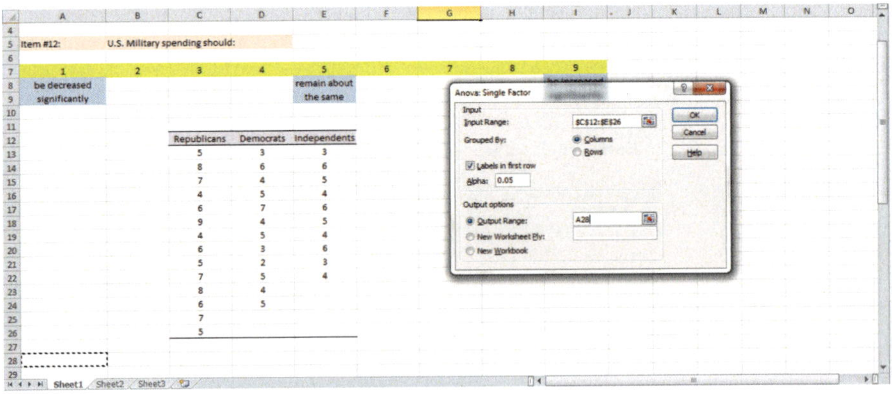

Fig. 8.3 Dialog Box for Anova: Single Factor Input/Output Range

OK

Next, format all decimal figures in the SUMMARY to two decimal places as this will make the printout much easier to read. Center all figures in their cells.

Save this file as: Military12

You should have generated the table given in Fig. 8.4.

	A	B	C	D	E	F	G
28	Anova: Single Factor						
29							
30	SUMMARY						
31	*Groups*	*Count*	*Sum*	*Average*	*Variance*		
32	Republicans	14	87	6.21	2.34		
33	Democrats	12	53	4.42	1.90		
34	Independents	10	46	4.60	1.38		
35							
36							
37	ANOVA						
38	*Source of Variation*	*SS*	*df*	*MS*	*F*	*P-value*	*F crit*
39	Between Groups	25.33	2	12.66	6.56	0.00	3.28
40	Within Groups	63.67	33	1.93			
41							
42	Total	89	35				
43							

Fig. 8.4 ANOVA Results for U.S. Military Spending

Print out both the data table and the ANOVA summary table so that all of this information fits onto one page. (Hint: Set the Page Layout / Fit to Scale to *70 % size*).

As a check on your analysis, you should have the following in these cells:

A28: Anova: Single Factor
D39: 12.66
G39: 3.28
D32: 6.21

Now, let's discuss how you should interpret this table:

8.2 How to Interpret the ANOVA Table Correctly

Objective: To interpret the ANOVA table correctly

ANOVA allows you to test for the differences between means when you have three or more groups of data. This ANOVA test is called the F-test statistic, and is typically identified with the letter: F.

The formula for the F-test is this:

F = Mean Square between groups (MS_b) divided by Mean Square within groups (MS_w)

$$F = MS_b / MS_w \tag{8.1}$$

The derivation and explanation of this formula is beyond the scope of this *Excel Guide*. In this *Excel Guide*, we are attempting to teach you *how to use Excel*, and we are not attempting to teach you the statistical theory that is behind the ANOVA formulas. For a detailed explanation of ANOVA, see Weiers (2011).

Note that cell D39 contains $MS_b = 12.66$, while cell D40 contains $MS_w = 1.93$.

When you divide these two figures using their cell references in Excel, you get the answer for the F-test of 6.56 which is in cell E39. Let's discuss now the meaning of the figure: $F = 6.56$.

In order to determine whether this figure for F of 6.56 indicates a significant difference between the means of the three groups, the first step is to write the null hypothesis and the research hypothesis for the three groups.

In statistics, the null hypothesis states that the population means of the three groups are equal, while the research hypothesis states that the population means of the three groups are not equal, and that there is, therefore, a significant difference between the population means of the three groups. Which of these two hypotheses should you accept based on the ANOVA results?

8.3 Using the Decision Rule for the ANOVA F-Test

To state the hypotheses, let's call Republicans as Group 1, Democrats as Group 2, and Independents as Group 3. The hypotheses would then be:

$H_0: \ \mu_1 = \mu_2 = \mu_3$
$H_1: \ \mu_1 \neq \mu_2 \neq \mu_3$

The answer to this question is analogous to the decision rule used in this book for both the one-group t-test and the two-group t-test. You will recall that this rule (see Sects. 4.1.6 and 5.1.8) was:

If the absolute value of t is less than the critical t, you accept the null hypothesis.

or

If the absolute value of t is greater than the critical t, you reject the null hypothesis, and accept the research hypothesis.

Now, here is the decision rule for ANOVA:

Objective: To learn the decision rule for the ANOVA F-test

The decision rule for the ANOVA F-test is the following:

If the value for F is less than the critical F-value, accept the null hypothesis.

or

If the value of F is greater than the critical F-value, reject the null hypothesis, and accept the research hypothesis.

Note that Excel tells you the critical F-value in cell G39: 3.28

Therefore, our decision rule for the three groups ANOVA test is this:

Since the value of F of 6.56 is greater than the critical F-value of 3.28, we reject the null hypothesis and accept the research hypothesis.

Therefore, our conclusion, in plain English, is:

There is a significant difference between the three groups of registered voters in Missouri in their attitudes toward the amount of U.S. military spending by the U.S. government.

Note that it is not necessary to take the absolute value of F of 6.56. The F-value can never be less than one, and so it can never be a negative value which requires us to take its absolute value in order to treat it as a positive value.

It is important to note that ANOVA tells us that there was a significant difference between the population means of the three groups, *but it does not tell us which pairs of groups were significantly different from each other.*

8.4 Testing the Difference Between Two Groups Using the ANOVA t-Test

To answer that question, we need to do a different test called the ANOVA t-test.

Objective: To test the difference between the means of two groups using an ANOVA t-test when the ANOVA results indicate a significant difference between the population means.

Since we have three groups of data (one group for each of the three groups of registered voters), we would have to perform three separate ANOVA t-tests to determine which pairs of groups were significantly different. This means that we would have to perform a separate ANOVA t-test for the following pairs of groups:

1. Republicans vs. Democrats
2. Republicans vs. Independents
3. Democrats vs. Independents

We will do just one of these pairs of tests, Republicans vs. Democrats, to illustrate the way to perform an ANOVA t-test comparing these two groups. The ANOVA t-test for the other two pairs of groups would be done in the same way.

8.4.1 Comparing Republicans vs. Democrats in Their Attitude Toward U.S. Military Spending Using the ANOVA t-Test

Objective: To compare Republicans vs. Democrats in their attitude toward U.S. Military spending using the ANOVA t-test.

The first step is to write the null hypothesis and the research hypothesis for these two groups.

For the ANOVA t-test, the null hypothesis is that the population means of the two groups are equal, while the research hypothesis is that the population means of the two groups are not equal (i.e., there is a significant difference between these two means). Since we are comparing Republicans (Group 1) vs. Democrats (Group 2), these hypotheses would be:

$H_0: \mu_1 = \mu_2$
$H_1: \mu_1 \neq \mu_2$

For Group 1 vs. Group 2, the formula for the ANOVA t-test is:

$$ANOVA\ t = \frac{\overline{X}_1 - \overline{X}_2}{s.e._{ANOVA}} \tag{8.2}$$

where

$$s.e._{ANOVA} = \sqrt{MS_w\left(\frac{1}{n_1} + \frac{1}{n_2}\right)} \tag{8.3}$$

The steps involved in computing this ANOVA t-test are:

1. Find the difference of the sample means for the two groups $(6.21 - 4.42 = 1.79)$.
2. Find $1/n_1 + 1/n_2$ (since both groups have a different number of people in them, this becomes: $1/14 + 1/12 = 0.07 + 0.08 = 0.15$)
3. Multiply MS_w times the answer for step 2 (1.93 x 0.15 = 0.29)
4. Take the square root of step 3 (SQRT(0.29) = 0.54)
5. Divide Step 1 by Step 4 to find ANOVA t (1.79/0.54 = 3.31)

Note: Since Excel computes all calculations to 16 decimal places, when you use your calculator for the above computations, your answer will be 3.31 in two decimal places, but Excel's answer of 3.29 will be more accurate because Excel's answer is always in 16 decimal places.

Now, what do we do with this ANOVA t-test result of 3.29? In order to interpret this value of 3.29 correctly, we need to determine the critical value of t for the ANOVA t-test. To do that, we need to find the degrees of freedom for the ANOVA t-test as follows:

8.4.1.1 Finding the Degrees of Freedom for the ANOVA t-Test

Objective: To find the degrees of freedom for the ANOVA t-test.

The degrees of freedom (df) for the ANOVA t-test is found as follows:

df = take the total sample size of all of the groups and subtract the numberof groups in your study ($n_{TOTAL} - k$ where k = the number of groups)

In our example, the total sample size of the three groups is 36 since there are 14 voters in Group1, 12 voters in Group 2, and 10 voters in Group 3, and since there are three groups, 36–3 gives a degrees of freedom for the ANOVA t-test of 33.

If you look up df = 33 in the t-table in Appendix E in the degrees of freedom column (df), which is the *second column on the left of this table*, you will find that the critical t-value is 2.035.

Important note: Be sure to use the degrees of freedom column (df) in Appendix E for the ANOVA t-test critical t value

8.4.1.2 Stating the Decision Rule for the ANOVA t-Test

Objective: To learn the decision rule for the ANOVA t-test

Interpreting the result of the ANOVA t-test follows the same decision rule that we used for both the one-group t-test (see Sect. 4.1.6) and the two-group t-test (see Sect. 5.1.8):

If the absolute value of t is less than the critical value of t, we accept the null hypothesis.

or

If the absolute value of t is greater than the critical value of t, we reject the null hypothesis and accept the research hypothesis.

Since we are using a type of t-test, we need to take the absolute value of t. Since the absolute value of 3.29 is greater than the critical t-value of 2.035, we reject the null hypothesis (that the population means of the two groups are equal) and accept the research hypothesis (that the population means of the two groups are significantly different from one another).

This means that our conclusion, in plain English, is as follows:

Republicans were significantly more willing than Democrats to increase U.S. Military spending (6.21 vs. 4.42).

Note that this difference in ratings of 1.79 points might not seem like much, but in practical terms, this means that the average ratings in the Republicans group are 40 % higher than the average ratings in the Democrats group. This, clearly, is an important and significant difference in attitude toward U.S. Military spending based on our hypothetical data.

8.4.1.3 Performing an ANOVA t-Test Using Excel Commands

Now, let's do these calculations for the ANOVA t-test using Excel with the file you created earlier in this chapter: Military12

A45: Republicans vs. Democrats
A47: 1/n Republicans + 1/n Democrats
A49: s.e. of Republicans vs. Democrats
A51: ANOVA t-test
C47: =(1/14 + 1/12)
C49: =SQRT(D40*C47)
C51: =(D32–D33)/C49

You should now have the following results in these cells when you round off all these figures in the ANOVA t-test to two decimal points.:

C47: 0.15
C49: 0.55
C51: 3.29

Re-save this final result under the file name: Military12

US. MILITARY SPENDING

Item #12: U.S. Military spending should:

1	2	3	4	5	6	7	8	9
be decreased significantly				remain about the same				be increased significantly

Republicans	Democrats	Independents
5	3	3
8	6	6
7	4	5
4	5	4
6	7	6
9	4	5
4	5	4
6	3	6
5	2	3
7	5	4
8	4	
6	5	
7		
5		

Anova: Single Factor

SUMMARY

Groups	Count	Sum	Average	Variance
Republicans	14	87	6.21	2.34
Democrats	12	53	4.42	1.90
Independents	10	46	4.60	1.38

ANOVA

Source of Variation	SS	df	MS	F	P-value	F crit
Between Groups	25.33	2	12.66	6.56	0.00	3.28
Within Groups	63.67	33	1.93			
Total	89	35				

Republicans vs. Democrats

1/n Republicans + 1/n Democrats	0.15
s.e. of Republicans vs. Democrats	0.55
ANOVA t-test	3.29

Fig. 8.5 Final Spreadsheet of U.S. Military Spending for Republicans vs. Democrats

Print out the resulting spreadsheet so that it fits onto one page like Fig. 8.5 (Hint: Reduce the Page Layout / Scale to Fit to *70 %*).

For a more detailed explanation of the ANOVA t-test, see Black (2010).

Important note: You are only allowed to perform an ANOVA t-test comparing the population means of two groups when the F-test produces a significant difference between the population means of all of the groups in your study.

It is improper to do any ANOVA t-test when the value of F is less than the critical value of F. Whenever F is less than the critical F, this means that there was no difference between the population means of the groups, and, therefore, that you cannot test to see if there is a difference between the means of any two groups since this would capitalize on chance differences between these two groups.

8.5 End-of-Chapter Practice Problems

1. Suppose you wanted to study the attitudes of different age groups of registered voters in Missouri toward the legal age for drinking alcohol in the state. Missouri currently has a legal age of 18 for drinking alcohol, but suppose you were asked to do a phone survey of registered voters in the state to see if different age groups felt differently about this question. Item #22 of the survey asked registered voters in Missouri about their attitude toward increasing the legal age for drinking alcohol from 18 to 21 using a 5-point Likert Scale (Strongly Agree to Strongly Disagree). You have decided to test your Excel skills on a small sample of randomly selected registered voters before you dive into the entire database of survey responses. The hypothetical data for this one item for five age groups appear in Fig. 8.6:

LEGAL AGE FOR DRINKING ALCOHOL IN MISSOURI

Question: Should the legal age for drinking alcohol in Missouri be increased from age 18 to age 21?

Item #22: The legal age for drinking alcohol in Missouri should be changed from 18 years to 21 years

1	2	3	4	5
Strongly disagree	Disagree	Undecided	Agree	Strongly Agree

		Age		
18 - 21	22 - 30	31 - 40	41 - 50	51+
5	4	2	3	1
4	3	3	5	3
3	2	4	4	4
4	3	3	5	2
5	2	5	4	4
2	4	4	3	3
1	3	3	2	2
4	5	5	5	1
3	4	4	4	2
5	3	5	5	3
4	4		3	5
5	3		4	4
4			4	2
3				1
				2

Fig. 8.6 Worksheet Data for Chap. 8: Practice Problem #1

(a) Enter these data on an Excel spreadsheet.
(b) Perform a *one-way ANOVA test* on these data, and show the resulting ANOVA table *underneath* the input data for the five age groups.
(c) If the F-value in the ANOVA table is significant, create an Excel formula to compute the ANOVA t-test comparing the average for Age 18–21 against Age 51+ and show the results below the ANOVA table on the spreadsheet (put the standard error and the ANOVA t-test value on separate lines of your spreadsheet, and use two decimal places for each value)
(d) Print out the resulting spreadsheet so that all of the information fits onto one page
(e) Save the spreadsheet as: Alcohol22

Let's call Age 18–21 Group 1, Age 22–30 Group 2, Age 31–40 Group 3, Age 41–50 Group 4, and Age 51+ Group 5.
Now, write the answers to the following questions using your Excel printout:

1. What are the null hypothesis and the research hypothesis for the ANOVA F-test?
2. What is MS_b on your Excel printout?
3. What is MS_w on your Excel printout?
4. Compute $F = MS_b / MS_w$ using your calculator.
5. What is the critical value of F on your Excel printout?
6. What is the result of the ANOVA F-test?
7. What is the conclusion of the ANOVA F-test in plain English?
8. If the ANOVA F-test produced a significant difference between the five age groups in their attitude toward the legal age for drinking alcohol,what is the null hypothesis and the research hypothesis for the ANOVA t-test comparing Age 18–21 versus Age 51 + ?
9. What is the mean (average) for Age 18–21 on your Excel printout?
10. What is the mean (average) for Age 51+ on your Excel printout?
11. What are the degrees of freedom (df) for the ANOVA t-test comparing Age 18–21 versus Age 51 + ?
12. What is the critical t value for this ANOVA t-test in Appendix E for these degrees of freedom?
13. Compute the $s.e._{ANOVA}$ using your calculator.
14. Compute the ANOVA t-test value comparing Age 18–21 versus Age 51 + using your calculator.
15. What is the *result* of the ANOVA t-test comparing Age 18–21 versus Age 51 + ?
16. What is the *conclusion* of the ANOVA t-test comparing Age 18–21 versus Age 51+ in plain English?

Note that since there are five age groups, you need to do ten ANOVA t-tests to determine what the significant differences are between the five groups. *Since you have just completed the ANOVA t-test comparing Age 18–21 versus Age 51+, you would have to do the ANOVA t-test next comparing the following nine groups in order to complete your ANOVA analysis for this one question:*

(a) Age 18–21 vs. Age 22–30
(b) Age 18–21 vs. Age 31–40
(c) Age 18–21 vs. Age 41–50
(d) Age 22–30 vs. Age 31–40
(e) Age 22–30 vs. Age 41–50
(f) Age 22–30 vs. Age 51+
(g) Age 31–40 vs. Age 41–50
(h) Age 31–40 vs. Age 51+
(i) Age 41–50 vs. Age 51+

2. Suppose that you wanted to determine the voting patterns of different age groups in the state of Illinois in the last Presidential Election. You have decided to take a random sample of the 102 counties in Illinois to study this question. You have dived into the data to determine the percent of registered voters who actually voted in this election in these counties, and broken down the data further to determine the percent of registered voters in each age group in these counties that voted in this election. You ran out of time to complete this analysis, but you wanted to test your Excel skills on a small sample of voters even though you had incomplete data on these counties. The hypothetical data for the voting patterns of the counties in which you selected are given in Fig. 8.7.

VOTING PATTERNS IN ILLINOIS IN THE LAST PRESIDENTIAL ELECTION

Question: Does age affect the percent of registered voters who voted?

	Age		
18 - 30	31 - 50	51 - 65	66 +
23	27	42	56
14	16	25	62
31	34	46	43
41	44	62	67
35	37	43	51
24	27	37	
21	25	34	
36	38		
24			

Fig. 8.7 Worksheet Data for Chap. 8: Practice Problem #2

(a) Enter these data on an Excel spreadsheet.
(b) Perform a *one-way ANOVA test* on these data, and show the resulting ANOVA table *underneath* the input data for the four age groups.

(c) If the F-value in the ANOVA table is significant, create an Excel formula to compute the ANOVA t-test comparing the voting patterns of Age 18–30 against the voting patterns of Age 51–65, and show the results below the ANOVA table on the spreadsheet (put the standard error and the ANOVA t-test value on separate lines of your spreadsheet, and use two decimal places for each value)
(d) Print out the resulting spreadsheet so that all of the information fits onto one page
(e) Save the spreadsheet as: Illinois3

Let's call Age 18–30 Group 1, Age 31–50 Group 2, Age 51–65 Group 3, and Age 66+ Group 4.
Now, write the answers to the following questions using your Excel printout:

1. What are the null hypothesis and the research hypothesis for the ANOVA F-test?
2. What is MS_b on your Excel printout?
3. What is MS_w on your Excel printout?
4. Compute $F = MS_b/MS_w$ using your calculator.
5. What is the critical value of F on your Excel printout?
6. What is the result of the ANOVA F-test?
7. What is the conclusion of the ANOVA F-test in plain English?
8. If the ANOVA F-test produced a significant difference between the four types of age groups in their voting patterns, what is the null hypothesis and the research hypothesis for the ANOVA t-test comparing Age 18–30 (Group 1) versus Age 51–65 (Group 3)?
9. What percent of registered voters voted in Age 18–30 on your Excel printout?
10. What percent of registered voters voted in Age 51–65 on your Excel printout?
11. What are the degrees of freedom (df) for the ANOVA t-test comparing Age 18–30 versus Age 51–65?
12. What is the critical t value for this ANOVA t-test in Appendix E for these degrees of freedom?
13. Compute the $s.e._{ANOVA}$ using your calculator for Age 18–30 versus Age 51–65.
14. Compute the ANOVA t-test value comparing Age 18–30 versus Age 51–65 using your calculator.
15. What is the *result* of the ANOVA t-test comparing Age 18–30 versus Age 51–65?
16. What is the *conclusion* of the ANOVA t-test comparing Age 18–30 versus Age 51–65 in plain English?

Important note: In order to perform a complete ANOVA analysis of these data, you would need to run ANOVA t-test comparing all six of the possible combinations of age groups, not just the one comparing Age 18–30 to Age 51–65. This means that you would need to compare the following age groups using an ANOVA t-test in addition to the one ANOVA t-test you performed above:

Age 18–30 vs. Age 31–50
Age 18–30 vs. Age 66+
Age 31–50 vs. Age 51–65
Age 31–50 vs. Age 66+
Age 51–65 vs. Age 66+

Once you have completed all six ANOVA t-test, you can then prepare a summary of all six of these tests in order to write a report that included testing all four age groups against one another.

3. Suppose that you have been hired as a research assistant by a political science professor who is interested in studying the relationship between political leaning (liberal, moderate, conservative) and the attitude of registered voters in Illinois toward U.S. defense spending using a phone survey. The professor has developed a survey in which Question #1 asks the respondent for his or her political leaning, and Question #3 asks the respondent for his or her attitude toward U.S. government spending on defense as a percent of the U.S. national budget. The hypothetical data from question #1 is used to generate the hypothetical data from Question #3 that is given in Fig. 8.8:

U.S. DEFENSE SPENDING AS A PERCENT OF THE NATIONAL BUDGET

Question: Does political leaning affect attitude toward U.S. defense spending?

Item #1: Do you consider yourself to be liberal, moderate, or conservative in your political views?

Item #3: What is your attitude toward U.S. government spending on defense as a percentage of the U.S. national budget?

1	2	3	4	5	6	7	8	9	10
It should be decreased				It should remain about the same					It should be increased

Political preference		
Liberal	Moderate	Conservative
3	4	4
5	5	7
4	6	6
6	4	8
7	5	5
5	3	9
4	7	10
3	6	7
2	4	5
1		6
		8

Fig. 8.8 Worksheet Data for Chap. 8: Practice Problem #3

(a) Enter these data on an Excel spreadsheet.
(b) Perform a *one-way ANOVA test* on these data, and show the resulting ANOVA table *underneath* the input data for the three types of political leaning.
(c) If the F-value in the ANOVA table is significant, create an Excel formula to compute the ANOVA t-test comparing the average for Moderates against the

average for Conservatives, and show the results below the ANOVA table on the spreadsheet (put the standard error and the ANOVA t-test value on separate lines of your spreadsheet, and use two decimal places for each value)

(d) Print out the resulting spreadsheet so that all of the information fits onto one page

(e) Save the spreadsheet as: Defense3

Now, write the answers to the following questions using your Excel printout:

1. What are the null hypothesis and the research hypothesis for the ANOVA F-test?
2. What is MS_b on your Excel printout?
3. What is MS_w on your Excel printout?
4. Compute $F = MS_b/MS_w$ using your calculator.
5. What is the critical value of F on your Excel printout?
6. What is the result of the ANOVA F-test?
7. What is the conclusion of the ANOVA F-test in plain English?
8. If the ANOVA F-test produced a significant difference between the three types of political leaning, what is the null hypothesis and the research hypothesis for the ANOVA t-test comparing Moderates versus Conservatives?
9. What is the mean (average) for Moderates on your Excel printout?
10. What is the mean (average) for Conservatives on your Excel printout?
11. What are the degrees of freedom (df) for the ANOVA t-test comparing Moderates versus Conservatives?
12. What is the critical t value for this ANOVA t-test in Appendix E for these degrees of freedom?
13. Compute the $s.e._{ANOVA}$ using your calculator for Moderates versus Conservatives.
14. Compute the ANOVA t-test value comparing Moderates versus Conservatives using your calculator.
15. What is the *result* of the ANOVA t-test comparing Moderates versus Conservatives?
16. What is the *conclusion* of the ANOVA t-test comparing Moderates versus Conservatives in plain English?

Important note: In order to perform a complete ANOVA analysis of these data, you would need to run ANOVA t-test comparing all three of the possible combinations of political leanings, not just the one comparing Moderates to Conservatives. This means that you would need to compare the following groups using an ANOVA t-test in addition to the one ANOVA t-test you performed above:

Liberals vs. Moderates

Liberals vs. Conservatives

Once you have completed all three ANOVA t-test, you can then prepare a summary of all three of these tests in order to write a report that included testing all three groups against one another.

References

Black, K. Business Statistics: For Contemporary Decision Making (6th ed.). Hoboken, NJ: John Wiley & Sons, Inc., 2010.

Weiers, R.M. Introduction to Business Statistics (7th ed.). Mason, OH: South-Western Cengage Learning, 2011.

Appendices

Appendix A: Answers to End-of-Chapter Practice Problems

© Springer International Publishing Switzerland 2016

T.J. Quirk, *Excel 2016 for Social Science Statistics*, Excel for Statistics,
DOI 10.1007/978-3-319-39711-5

Chapter 1: Practice Problem #1 Answer (see Fig. A.1)

Chapter 1: Practice Problem # 1 Answer							
POLITICAL SCIENCE SURVEY OF U.S. - CHINESE RELATIONS							
Item #8:	"The Chinese leaders are basically trying to get along with the U.S."						
	1	2	3	4	5	6	7
	strongly disagree			undecided			strongly agree
				Data			
				6			
				4			
				5		n	17
				3			
				2			
				7		mean	3.88
				5			
				3			
				4		stdev	1.65
				2			
				4			
				6		s.e.	0.40
				3			
				5			
				4			
				2			
				1			

Fig. A.1 Answer to Chap. 1: Practice Problem #1

Chapter 1: Practice Problem #2 Answer (see Fig. A.2)

Chapter 1: Practice Problem #2 Answer

HUMAN RESOURCES MORALE SURVEY

Item #21: **"Management is doing a good job of keeping employee morale at a high level."**

1 Disagree	2	3	4	5	6	7 Agree

Rating
3
6
5
7
2
3
6
5
4
7
6
1
3
2
4
5
6
4
5
3
6
4
7

n	23
Mean	4.52
STDEV	1.73
s.e.	0.36

Fig. A.2 Answer to Chap. 1: Practice Problem #2

Chapter 1: Practice Problem #3 Answer (see Fig. A.3)

Chapter 1: Practice Problem #3: Answer

Deer Creek Elementary School

5th grade science test

Chapter 8 (15 items)		
12		
15	n	16
13		
8		
10	MEAN	11.750
12		
13		
12	STDEV	2.910
9		
4		
11	s.e.	0.727
15		
13		
15		
12		
14		

Fig. A.3 Answer to Chap. 1: Practice Problem #3

Chapter 2: Practice Problem #1 Answer (see Fig. A.4)

Fig. A.4 Answer to Chap. 2: Practice Problem #1

Chapter 2: Practice Problem #1 Answer

FRAME NUMBERS	Duplicate frame numbers	RANDOM NO.
1	44	0.477
2	33	0.692
3	38	0.838
4	43	0.834
5	13	0.869
6	10	0.991
7	50	0.993
8	1	0.955
9	48	0.053
10	61	0.652
11	4	0.921
12	22	0.907
13	40	0.638
14	37	0.383
15	35	0.911
16	60	0.436
17	59	0.103
18	7	0.148
19	17	0.230
20	30	0.083
21	29	0.850
22	[illegible]	[illegible]69
[illegible]	42	0.[illegible]
55	11	0.190
56	56	0.472
57	57	0.456
58	54	0.818
59	9	0.516
60	51	0.304
61	39	0.985
62	53	0.788
63	26	0.398

Chapter 2: Practice Problem #2 Answer (see Fig. A.5)

Fig. A.5 Answer to Chap. 2: Practice Problem #2

Chapter 2: Practice Problem #2 Answer

FRAME NO.	Duplicate frame no.	Random number
1	45	0.985
2	102	0.846
3	16	0.133
4	8	0.007
5	109	0.869
6	64	0.525
7	37	0.241
8	31	0.851
9	27	0.393
10	76	0.362
11	9	0.859
12	70	0.380
13	13	0.328
14	32	0.593
15	56	0.443
16	46	0.026
17	3	0.215
18	98	0.683
19	10	0.827
20	100	0.739
21	29	0.721
[illegible]	39	
		0.805
98	59	0.659
99	2	0.972
100	35	0.683
101	20	0.618
102	73	0.015
103	11	0.947
104	24	0.096
105	82	0.025
106	5	0.304
107	17	0.592
108	34	0.409
109	104	0.252
110	51	0.223
111	6	0.961
112	84	0.296
113	96	0.908
114	67	0.236

Chapter 2: Practice Problem #3 Answer (see Fig. A.6)

Fig. A.6 Answer to Chap. 2: Practice Problem #3

Chapter 2: Practice Problem #3 Answer

FRAME NUMBERS	Duplicate frame numbers	Random number
1	47	0.347
2	68	0.739
3	15	0.535
4	69	0.925
5	67	0.584
6	38	0.073
7	43	0.172
8	50	0.265
9	65	0.384
10	40	0.812
11	57	0.739
12	37	0.419
13	22	0.112
14	3	0.792
15	17	0.554
16	60	0.149
17	5	0.402
18	29	0.654
19	74	0.870
20	72	0.902
21	14	0.835
22	41	0.055
23	53	0.238
24	9	0.488
25	19	0.019
26	33	0.371
27	71	0.995
28	49	0.943
29	1	0.865
30	16	0.272
31		433
	28	0.
53	4	0.312
54	31	0.402
55	32	0.998
56	34	0.368
57	58	0.777
58	20	0.588
59	21	0.971
60	26	0.821
61	36	0.170
62	70	0.346
63	39	0.303
64	2	0.275
65	54	0.982
66	44	0.020
67	25	0.367
68	61	0.523
69	23	0.759
70	27	0.570
71	46	0.169
72	35	0.954
73	11	0.759
74	7	0.636
75	12	0.106
76	30	0.213

Chapter 3: Practice Problem #1 Answer (see Fig. A.7)

Chapter 3: Practice Problem #1: Answer

HEALTH CARE REFORM SURVEY

Item #10: "What type of health care program should be adopted in the U.S.?"

0	1	2	3	4	5	6	7	8	9	10
favor a private medical insurance plan					undecided			favor a government-funded medical insurance plan		

Data		
3		
6	Null hypothesis:	μ = 5
4		
5		
7	Research hypothesis:	μ ≠ 5
9		
1		
4	n	17
2		
6		
4	Mean	5.47
7		
9		
10	STDEV	2.60
5		
3		
8	s.e.	0.63

95% confidence interval

lower limit	4.13
upper limit	6.81

4.13 - - - - - - - - - - - 5 - - - - - - - -5.47 - - - - - - - - - - - 6.81

lower limit	Ref. value	Mean	upper limit

Result: Since the reference value pf 5 is inside the confidence interval, we accept the null hypothesis

Conclusion: Registered voters in Illinois were undecided as to whether they favored a government-funded medical insurance plan or a private medical insurance plan

Fig. A.7 Answer to Chap. 3: Practice Problem #1

Chapter 3: Practice Problem #2 Answer (see Fig. A.8)

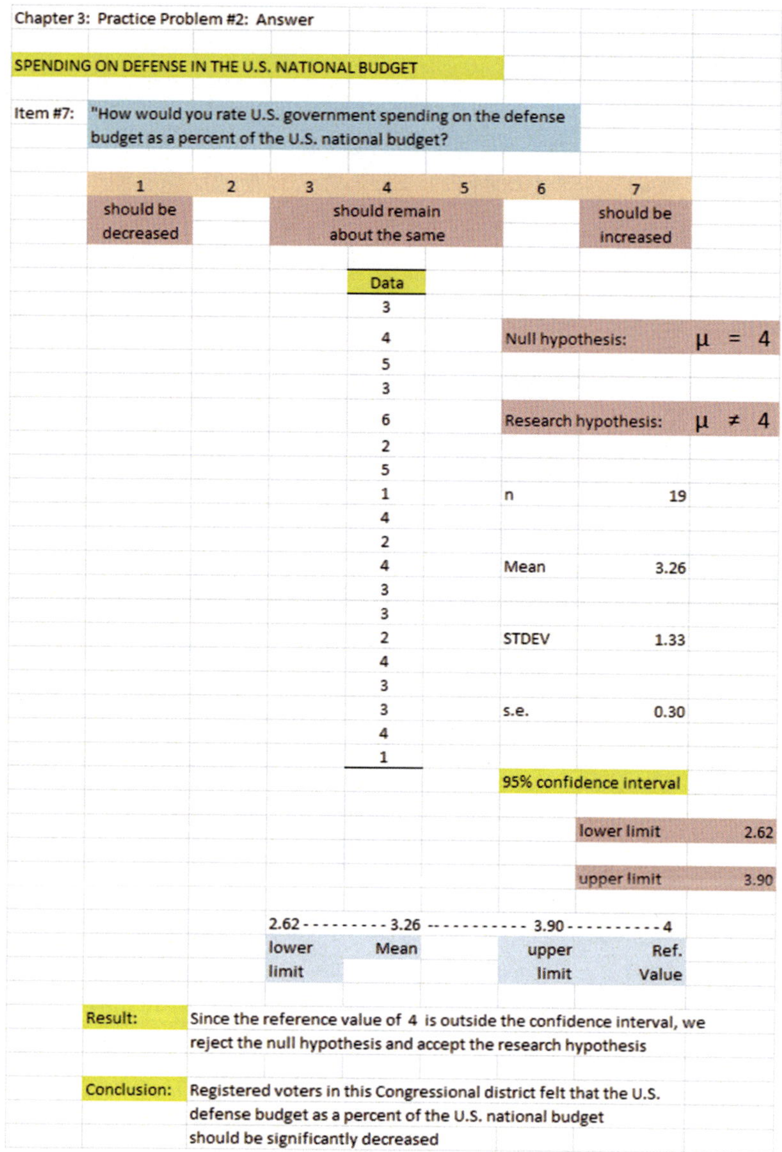

Fig. A.8 Answer to Chap. 3: Practice Problem #2

Chapter 3: Practice Problem #3 Answer (see Fig. A.9)

Chapter 3: Practice Problem #3: Answer

WELCH'S 100% GRAPE JUICE

Research question: "Does the average can of Welch's 100% Grape Juice produced today contain 163 ml of grape juice?"

ml			
165	Null hypothesis:		μ = 163 ml
158			
163			
159	Research hypothesis:		μ ≠ 163 ml
154			
157			
159		n	13
161			
164		Mean	159.62
154			
157		STDEV	3.59
161			
163		s.e.	1.00

95% confidence interval

lower limit	157.44
upper limit	161.79

157.44 - - - - - - -159.62 - - - - - - - - - 161.79 - - - - - - 163

lower limit	Mean	upper limit	Ref. Value

Result: Since the reference value of 163 ml is outside of the confidence interval, we reject the null hypothesis and accept the research hypothesis

Conclusion: Cans of 100% Grape Juice produced today contained significantly less than 163 ml , and it was probably closer to 160 ml

Fig. A.9 Answer to Chap. 3: Practice Problem #3

Chapter 4: Practice Problem #1 Answer (see Fig. A.10)

Chapter 4: Practice Problem #1: Answer						
LIBERAL-CONSERVATIVISM OF COLLEGE STUDENTS						
Item #12:	"How would you rate your political ideology?"					
1	2	3	4	5	6	7
extermely liberal			moderate			extermely conservative
		Data				
		2				
		6		Null hypothesis:		μ = 4
		3				
		4				
		7		Research hypothesis:		μ ≠ 4
		5				
		3				
		5		n	20	
		4				
		2		mean	3.95	
		1				
		3		STDEV	1.76	
		4				
		2		s.e.	0.39	
		5				
		6				
		3		t-test	-0.13	
		5				
		2				
		7		critical t	2.093	
	Result:	Since the absolute value of −0.13 is less than the critical t				
		we accept the null hypothesis				
	Conclusion:	Undergraduates at this university do not consider themsel				
		either liberal or conservative, but moderate in their politi				

Fig. A.10 Answer to Chap. 4: Practice Problem #1

Chapter 4: Practice Problem #2 Answer (see Fig. A.11)

Chapter 4: Practice Problem #2: Answer

JOB TRAINING PROGRAM SURVEY

Item #18: "How would you rate the support and guidance you received in this program in securing employment?"

1	2	3	4	5
poor		satisfactory		excellent

Data		
3		
5	Null hypothesis:	$\mu = 3$
2		
4		
3	Research hypothesis:	$\mu \neq 3$
5		
4		
5	n	19
3		
4	mean	3.79
2		
5	STDEV	1.23
1		
5	s.e.	0.28
4		
5	critical t	2.101
3		
5	t-test	2.80
4		

Result: Since the absolute value of 2.80 is greater than the critical t of 2.101, we reject the null hypothesis and accept the research hypothesis

Conclusion: Participants in the Job Training Program rated the support and guidance they received in this program in securing employment as significantly positive (NOTE: In English, it does not make sense to say that something was "significantly excellent," so using the word "positive" here is a much clearer way to summarize the conclusion)

Fig. A.11 Answer to Chap. 4: Practice Problem #2

Chapter 4: Practice Problem #3 Answer (see Fig. A.12)

Chapter 4: Practice Problem #3 Answer

MISSOURI BOTANICAL GARDEN

VISITOR SURVEY

Item #10: "How would you rate the helpfulness of The Garden staff?"

1	2	3	4	5	6	7	8	9
poor						6.57 Mean		excellent

Results
8
6
5
7
9
5
6
4
8
7
6
8
6
7
9
7
6
3
8
7
6

Null hypothesis: $\mu = 5$

Research hypothesis $\mu \neq 5$

n	21
Mean	6.57
STDEV	1.54
s.e.	0.34
critical t	2.086
t-test	4.69

Result: Since the absolute value of 4.69 is greater than the critical value of 2.086, we reject the null hypothesis and accept the research hypothesis

Conclusion: Visitors to the Missouri Botanical Garden rated the helpfulness of The Garden staff as significantly positive.

Fig. A.12 Answer to Chap. 4: Practice Problem #3

Chapter 5: Practice Problem #1 Answer (see Fig. A.13)

Chapter 5: Practice Problem #1: Answer

ITBS TEST-SCORE EFFECTS OF SCHOOL VOUCHERS

Group	n	Mean	STDEV
1 Experimental	52	6.2	0.96
2 Control	57	5.3	1.12

Null hypothesis:	$\mu_1 = \mu_2$
Research hypothesis:	$\mu_1 \neq \mu_2$
STDEV1 squared / n1	0.02
STDEV2 squared / n2	0.02
B14 + B16	0.04
s.e.	0.20
critical t	1.96
t-test	4.52

Result: Since the absolute value of 4.52 is greater t
of 1.96, we reject the null hypothesis and ac
hypothesis

Conclusion: The experimental group had significantly hi
ITBS than the control group (6.2 vs. 5.3)

Fig. A.13 Answer to Chap. 5: Practice Problem #1

Chapter 5: Practice Problem #2 Answer (see Fig. A.14)

Chapter 5: Practice Problem #2: Answer

Item: "How interested are you in learning more about how life insurance can provide income for retirement?"

1	2	3	4	5	6	7
Not at all interested		3.44 Women		5.16 Men		Very Interested

Ad: Male model

Men	Women
5	3
6	4
4	6
7	5
5	2
6	3
5	1
4	3
3	2
6	4
7	3
5	5
6	6
4	3
7	4
5	2
4	5
6	3
3	4
7	5
5	4
6	3
2	2
6	4
1	3
7	5
6	1
5	3
4	2
6	3
5	2
7	5
	3
	4

Null hypothesis: $\mu_1 = \mu_2$

Research hypothesis: $\mu_1 \neq \mu_2$

Group	n	mean	STDEV
1 Men	32	5.16	1.51
2 Women	34	3.44	1.31

STDEV1 squared / n1 0.07

STDEV2 squared / n2 0.05

s.e. 0.35

ctitical t 1.96
(df = n1 + n2 -2 = 64)

t-test 4.93

Result: Since the absolute value of 4.93 is greater than the critical t of 1.96, we reject the null hypothesis and accept the research hypothesis

Conclusion: Adult me (ages 25-39) were significantly more interested than adult women (ages 25-39) in learning more about how life insurance can provide income for retirement when a male model was used in the ad (5.16 vs. 3.44)

Fig. A.14 Answer to Chap. 5: Practice Problem #2

Chapter 5: Practice Problem #3 Answer (see Fig. A.15)

Chapter 5: Practice Problem #3: Answer

EFFECT OF POLITICAL PREFERENCE ON ATTITUDE TOWARD GUN CONTROL

Null hypothesis: $\mu_1 = \mu_2$

Research hypothesis: $\mu_1 \neq \mu_2$

Item #13: Buying a gun should be:

1	2	3	4	5
a lot easier than it is now	somewhat easier than it is now	about the same as it is now	somewhat more difficult than it is now	a lot more difficult than it is now

Republicans	Democrats
1	2
2	5
4	4
3	3
5	5
3	4
2	3
4	4
1	5
2	4
3	
2	

Group	n	Mean	STDEV
1 Republicans	12	2.67	1.23
2 Democrats	10	3.90	0.99

(n1 - 1) x STDEV1 squared	16.67
(n2 - 1) x STDEV2 squared	8.90
n1 + n2 - 2	20
1/n1 + 1/n2	0.18
s.e.	0.48
critical t	2.086
t-test	-2.55

Result: Since the absolute value of − 2.55 is greater than the critical t of 2.086, we reject the null hypothesis and accept the research hypothesis

Conclusion: Democrats want it to be significantly more difficult to buy a gun than Republicans (3.90 vs. 2.67)

Fig. A.15 Answer to Chap. 5: Practice Problem #3

Chapter 6: Practice Problem #1 Answer (see Fig. A.16)

Chapter 6: Practice Problem #1: Answer

RELATIONSHIP BETWEEN COLLEGE DEGREES AND % OF WOMEN IN STATE LEGISLATORS

X = Percent of college degrees held for public high school graduates age 25+ by state

Y = Percent of state legislators who are women

% college degrees	% female state legislators
11	9
14	18
21	35
17	31
19	36
21	28
14	18
22	40
13	18
16	12
18	15
19	12

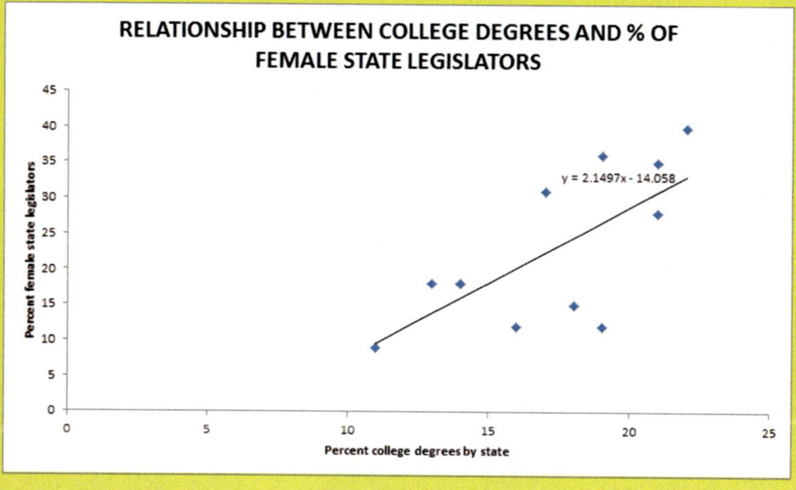

SUMMARY OUTPUT

Regression Statistics	
Multiple R	0.71
R Square	0.50
Adjusted R Square	0.45
Standard Error	7.96
Observations	12

ANOVA

	df	SS	MS	F	Significance F
Regression	1	632.74	632.74	9.98	0.01
Residual	10	633.93	63.39		
Total	11	1266.67			

	Coefficients	Standard Error	t Stat	P-value	Lower 95%
Intercept	-14.06	11.85	-1.19	0.26	-40.46
X Variable 1	2.15	0.68	3.16	0.01	0.63

Fig. A.16 Answer to Chap. 6: Practice Problem #1

Chapter 6: Practice Problem #1 (continued)

1. $a = \text{y-intercept} = -14.06$
2. $b = \text{slope} = 2.15$
3. $Y = a + bX$

   $Y = -14.06 + 2.15X$
4. $Y = -14.06 + 2.15(20)$

   $Y = -14.06 + 43$

   $Y = 28.94$, or about 29 percent

Chapter 6: Practice Problem #2 Answer (see Fig. A.17)

Chapter 6: Practice Problem #2 Answer

RELATIONSHIP BETWEEN DIET AND WEIGHT LOSS

ADULT WOMEN AGES 30-40

DIET (calories allowed per day)	WEIGHT LOSS (kg)
900	16.0
1050	12.0
1150	8.0
1275	6.0
1420	3.0
1530	5.5
1610	9.5
1710	2.5
1820	6.0
1875	9.0
1930	6.0
2100	3.0

correlation	-0.64

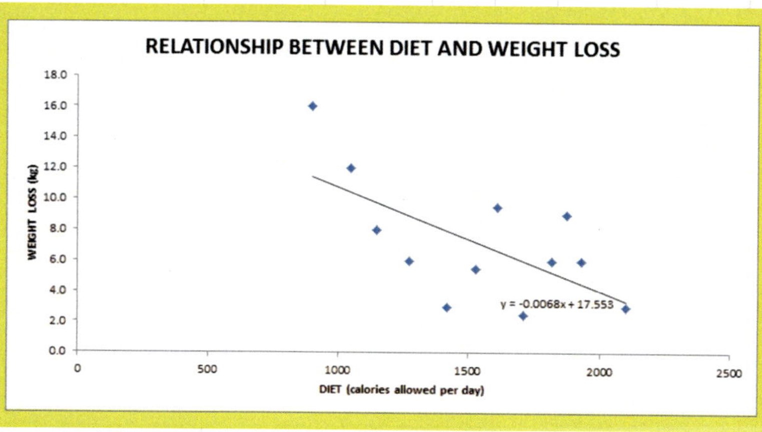

SUMMARY OUTPUT

Regression Statistics	
Multiple R	0.64
R Square	0.413
Adjusted R Square	0.354
Standard Error	3.198
Observations	12

ANOVA

	df	*SS*	*MS*	*F*	*Significance F*
Regression	1	71.946	71.946	7.034	0.024
Residual	10	102.284	10.228		
Total	11	174.229			

	Coefficients	*Standard Error*	*t Stat*	*P-value*	*Lower 95%*	*Upper 95%*
Intercept	17.553	4.008	4.379	0.001	8.622	26.483
X Variable 1	-0.007	0.003	-2.652	0.024	-0.012	-0.001

Fig. A.17 Answer to Chap. 6: Practice Problem #2

Chapter 6: Practice Problem #2 (continued)

1. $r = -0.64$ (note the negative correlation!)
2. $a = \text{y-intercept} = 17.553$
3. $b = \text{slope} = -0.007$ (note the minus sign as the slope is negative)
4. $Y = a + b\,X$
   $Y = 17.553 - 0.007\,X$
5. $Y = 17.553 - 0.007(1,500)$
   $Y = 17.553 - 10.5$
   $Y = 7.1\,\text{kg}$

Chapter 6: Practice Problem #3 Answer (see Fig. A.18)

Chapter 6: Practice Problem #3: Answer

RELATIONSHIP BETWEEN STATE POPULATION DENSITY AND TRAFFIC FATALITIES

X = State population density per square mile

Y = Motor vehicle fatalities within 30 days of the accident

Population density	Motor vehicle fatalities
17	65
25	123
80	612
124	136
724	3400
653	1800
541	613
376	512
412	316
114	363
52	163
91	25
correlation	0.80

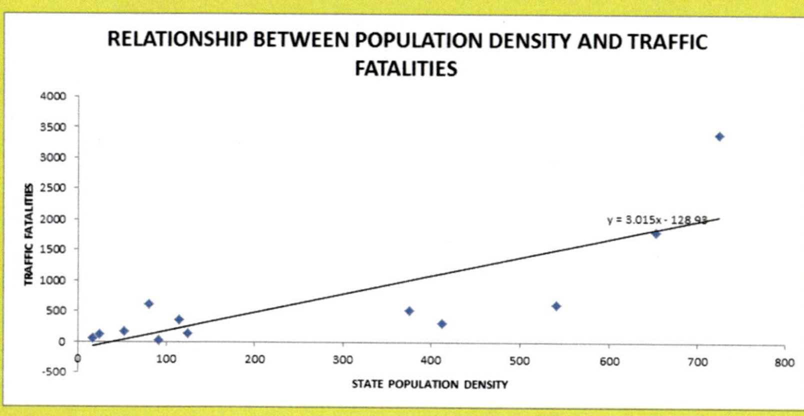

SUMMARY OUTPUT

Regression Statistics	
Multiple R	0.80
R Square	0.64
Adjusted R Square	0.60
Standard Error	620.51
Observations	12

ANOVA

	df	*SS*	*MS*	*F*	*Significance F*
Regression	1	6753383.41	6753383.41	17.54	0.001863799
Residual	10	3850377.26	385037.73		
Total	11	10603760.67			

	Coefficients	*Standard Error*	*t Stat*	*P-value*	*Lower 95%*
Intercept	-128.93	262.96	-0.49	0.63	-714.85
X Variable 1	3.02	0.72	4.19	0.00	1.41

Fig. A.18 Answer to Chap. 6: Practice Problem #3

Chapter 6: Practice Problem #3 (continued)

1. $r = .80$
2. $a = \text{y-intercept} = -128.93$
3. $b = \text{slope} = 3.02$
4. $Y = a + b\,X$
   $Y = -128.93 + 3.02\,X$
5. $Y = -128.93 + 3.02(670)$
   $Y = -128.93 + 2023.40$
   $Y = 1894.47$
   $Y = 1895$ traffic fatalities

Chapter 7: Practice Problem #1 Answer (see Fig. A.19)

Chapter 7: Practice Problem #1: Answer

MOTOR VEHICLE FATALITIES IN TERMS OF POPULATION DENSITY AND AGE OF RESIDENTS

Question: What is the relationship between the population density of a state, the percent of that state's population ages 18-24, and motor vehicle fatalities in that state?

	Y	X1	X2
State	Motor vehicle fatalities (per 100,000 population)	Population density (per square mile)	Percent of population (ages 18 - 24)
1	18	80	5
2	6	120	3
3	5	200	4
4	7	220	4
5	8	175	6
6	7	195	6
7	12	50	8
8	14	70	9
9	8	185	4
10	6	190	6

SUMMARY OUTPUT

Regression Statistics	
Multiple R	0.79
R Square	0.62
Adjusted R Square	0.51
Standard Error	2.94
Observations	10

ANOVA

	df	*SS*	*MS*
Regression	2	98.19	49.10
Residual	7	60.71	8.67
Total	9	158.90	

	Coefficients	*Standard Error*	*t Stat*
Intercept	15.49	5.60	2.77
Population density	-0.05	0.02	-2.62
% of population 18 - 24	0.18	0.62	0.29

	Motor vehicle fatalities (per 100,000 population)	*Population density (per square mile)*	*Percent of population (ages 18 - 24)*
Motor vehicle fatalities	1		
Population density	-0.78	1	
% of population 18 - 24	0.49	-0.56	1

Fig. A.19 Answer to Chap. 7: Practice Problem #1

Chapter 7: Practice Problem #1 (continued)

1. Multiple correlation = .79
2. y-intercept = 15.49
3. −0.05
4. 0.18
5. $Y = a + b_1 X_1 + b_2 X_2$
   $Y = 15.49 - 0.05 X_1 + 0.18 X_2$
6. $Y = 15.49 - 0.05(220) + 0.18(8)$
   $Y = 15.49 - 11 + 1.44$
   $Y = 5.93$
   $Y = \text{About 6 fatalities per 100,000 population}$
7. −.78
8. .49
9. −.56
10. Population density is the better predictor of Motor vehicle fatalities with a correlation of −.78.
11. The two predictors combined predict Motor vehicle fatalities at +.79, and this is only very slightly better than the better single predictor's correlation of −.78.

Chapter 7: Practice Problem #2 Answer (see Fig. A.20)

GRADUATE RECORD EXAMINATIONS

How well does the GRE and the GRE subject area test in Psychology predict
GPA at the end of the first year of a Masters' program in Psychology?

FIRST-YEAR GPA	GRE VERBAL	GRE QUANTITATIVE	GRE WRITING	GRE PSYCHOLOGY
3.25	160	161	5	650
3.42	156	158	4	600
2.85	156	157	2	500
2.65	154	153	1	510
3.65	166	166	6	630
3.16	159	160	3	550
3.56	166	163	4	610
2.35	155	154	2	430
2.86	153	154	3	450
2.95	158	157	4	550
3.15	158	159	4	580
3.45	160	160	5	620

SUMMARY OUTPUT

Regression Statistics	
Multiple R	0.93
R Square	0.868
Adjusted R Square	0.792
Standard Error	0.178
Observations	12

ANOVA

	df	*SS*	*MS*	*F*	*Significance F*
Regression	4	1.453	0.363	11.491	0.003
Residual	7	0.221	0.032		
Total	11	1.674			

	Coefficients	*Standard Error*	*t Stat*	*P-value*	*Lower 95%*
Intercept	-5.2698	4.173	-1.263	0.247	-15.138
GRE VERBAL	-0.0174	0.041	-0.424	0.684	-0.115
GRE QUANTITATIVE	0.0614	0.057	1.070	0.320	-0.074
GRE WRITING	0.0332	0.075	0.444	0.670	-0.144
GRE PSYCHOLOGY	0.0023	0.001	1.592	0.155	-0.001

	FIRST-YEAR GPA	*GRE VERBAL*	*GRE QUANTITATIVE*	*GRE WRITING*	*GRE MATH*
FIRST-YEAR GPA	1				
GRE VERBAL	0.79	1			
GRE QUANTITATIVE	0.88	0.94	1		
GRE WRITING	0.83	0.72	0.83	1	
GRE PSYCHOLOGY	0.89	0.74	0.83	0.81	1

Fig. A.20 Answer to Chap. 7: Practice Problem #2

Chapter 7: Practice Problem #2 (continued)

1. $R_{xy} = +0.93$
2. y-intercept $= -5.2698$
3. -0.0174
4. 0.0614
5. 0.0332
6. 0.0023
7. $Y = a + b_1X_1 + b_2X_2 + b_3X_3 + b_4X_4$
   $Y = -5.2698 - 0.0174X_1 + 0.0614X_2 + 0.0332X_3 + 0.0023X_4$

8. $Y = -5.2698 - 0.0174(160) + 0.0614(150) + 0.0332(3) + 0.0023(610)$
   $Y = -5.2698 - 2.78 + 9.21 + 0.10 + 1.40$
   $Y = 10.71 - 8.05$
   $Y = 2.66$
9. +0.79
10. +0.88
11. +0.83
12. +0.89
13. +0.72
14. +0.74
15. The best predictor of First-year GPA was the Psychology exam with a correlation $r = 0.89$
16. The four predictors combined predict First-year GPA with a correlation $R_{xy} = 0.93$ which is much better than the best single predictor by itself

Chapter 7: Practice Problem #3 Answer (see Fig. A.21)

Chapter 7: Practice Problem #3 Answer

	Y	X_1	X_2	X_3
Store ID	Annual Sales ($000)	Average Daily Traffic	Population (2-mile radius)	Average Income in Area
1	1,121	61,655	17,880	$28,991
2	766	35,236	13,742	$14,731
3	595	35,403	19,741	$8,114
4	899	52,832	23,246	$15,324
5	915	40,809	24,485	$11,438
6	782	40,820	20,410	$11,730
7	833	49,147	28,997	$10,589
8	571	24,953	9,981	$10,706
9	692	40,828	8,982	$23,591
10	1,005	39,195	18,814	$15,703
11	589	34,574	16,941	$9,015
12	671	26,639	13,319	$10,065
13	903	55,083	21,482	$17,365
14	703	37,892	26,524	$7,532
15	556	24,019	14,412	$6,950
16	657	27,791	13,896	$9,855
17	1,209	53,438	22,444	$21,589
18	997	54,835	18,096	$22,659
19	844	32,919	16,458	$12,660
20	883	29,139	16,609	$11,618

SUMMARY OUTPUT

Regression Statistics	
Multiple R	0.91
R Square	0.836
Adjusted R Square	0.806
Standard Error	81.603
Observations	20

ANOVA

	df	*SS*	*MS*	*F*	*Significance F*
Regression	3	544502.827	181500.942	27.256	1.58756E-06
Residual	16	106544.123	6659.008		
Total	19	651046.950			

	Coefficients	*Standard Error*	*t Stat*	*P-value*	*Lower 95%*	*Upper 95%*	*Lower 95.0%*	*Upper 95.0%*
Intercept	60.07	91.755	0.655	0.522	-134.440	254.585	-134.440	254.585
Average Daily Traffic	-0.02	0.006	-2.682	0.016	-0.030	-0.004	-0.030	-0.004
Population (2-mile radius)	0.04	0.009	4.401	0.000	0.021	0.060	0.021	0.060
Average Income in Area	0.05	0.010	4.898	0.000	0.028	0.071	0.028	0.071

	Annual Sales ($000)	*Average Daily Traffic*	*Population (2-mile radius)*	*Average Income in Area*
Annual Sales ($000)	1			
Average Daily Traffic	0.77	1		
Population (2-mile radius)	0.42	0.53	1	
Average Income in Area	0.72	0.74	-0.11	1

Fig. A.21 Answer to Chap. 7: Practice Problem #3

Chapter 7: Practice Problem #3 (continued)

1. $R_{xy} = 0.91$
2. y-intercept = 60.07
3. Average Daily Traffic = −0.02
4. Population = 0.04
5. Average Income = 0.05
6. $Y = a + b_1X_1 + b_2X_2 + b_3X_3$
   $Y = 60.07 - 0.02X_1 + 0.04X_2 + 0.05X_3$
7. $Y = 60.07 - 0.02(42,000) + 0.04(23,000) + 0.05(22,000)$
   $Y = 60.07 - 840 + 920 + 1,100$
   $Y = 1240.07$
   $Y = \$1,240,000$ or $\$1.24$ million
8. +0.77
9. +0.42
10. +0.72
11. +0.53
12. −0.11
13. Average Daily Traffic is the best predictor of Annual Sales because it has a correlation of +0.77 with Annual Sales, and the other two predictors have a correlation that is smaller than 0.77 (0.72 and 0.42).
14. The three predictors combined predict Annual Sales at +0.91 which is much better than the best single predictor by itself of 0.77.

Chapter 8: Practice Problem #1 Answer (see Fig. A.22)

Chapter 8: Practice Problem #1: Answer

LEGAL AGE FOR DRINKING ALCOHOL IN MISSOURI

Question: Should the legal age for drinking alcohol in Missouri be increased from age 18 to age 21?

Item #22: The legal age for drinking alcohol in Missouri should be changed from 18 years to 21 years

1	2	3	4	5
Strongly disagree	Disagree	Undecided	Agree	Strongly Agree

		Age		
18 - 21	22 - 30	31 - 40	41 - 50	51+
5	4	2	3	1
4	3	3	5	3
3	2	4	4	4
4	3	3	5	2
5	2	5	4	4
2	4	4	3	3
1	3	3	2	2
4	5	5	5	1
3	4	4	4	2
5	3	5	5	3
4	4		3	5
5	3		4	4
4			4	2
3				1
				2

Anova: Single Factor

Null hypothesis: $\mu_1 = \mu_2 = \mu_3 = \mu_4 = \mu_5$

Research hypothesis: $\mu_1 \neq \mu_2 \neq \mu_3 \neq \mu_4 \neq \mu_5$

SUMMARY

Groups	*Count*	*Sum*	*Average*	*Variance*
18 - 21	14	52	3.71	1.45
22 - 30	12	40	3.33	0.79
31 - 40	10	38	3.80	1.07
41 - 50	13	51	3.92	0.91
51+	15	39	2.60	1.54

ANOVA

Source of Variation	*SS*	*df*	*MS*	*F*	*P-value*	*F crit*
Between Groups	16.10	4	4.03	3.41	0.014	2.53
Within Groups	69.65	59	1.18			
Total	85.75	63				

Age 18 - 21 vs. Age 51 +

1/age 18-21 + 1/age 51+	0.14
s.e.	0.40
ANOVA t	2.76

Fig. A.22 Answer to Chap. 8: Practice Problem #1

Chapter 8: Practice Problem #1 (continued)

1. $H_0 : \mu_1 = \mu_2 = \mu_3 = \mu_4 = \mu_5$
   $H_1 : \mu_1 \neq \mu_2 \neq \mu_3 \neq \mu_4 \neq \mu_5$
2. 4.03
3. 1.18
4. $F = 3.42$
5. 2.53
6. Result: Since 3.42 is greater than 2.53, we reject the null hypothesis and accept the research hypothesis
7. There was a significant difference between age groups in terms of their attitude toward the legal age for drinking alcohol in Missouri.

   AGE 18–21 vs. AGE 51+

8. $H_0 : \mu_1 = \mu_5$
   $H_1 : \mu_1 \neq \mu_5$
9. 3.71
10. 2.60
11. $64 - 5 = 59$
12. critical $t = 1.96$
13. s.e. $= 0.41$
14. ANOVA $t = 2.71$
15. Result: Since the absolute value of 2.71 is greater than 1.96, we reject the null hypothesis and accept the research hypothesis
16. Conclusion: Age group 51+ was significantly more likely to disagree that the legal age for drinking alcohol in Missouri should be changed from 18 to 21 years than the age group 18–21 (2.60 vs. 3.71)

Chapter 8: Practice Problem #2 Answer (see Fig. A.23)

Chapter 8: Practice Problem #2: Answer

VOTING PATTERNS IN ILLINOIS IN THE LAST PRESIDENTIAL ELECTION

Question: Does age affect the percent of registered voters who voted?

Age			
18 - 30	31 - 50	51 - 65	66 +
23	27	42	56
14	16	25	62
31	34	46	43
41	44	62	67
35	37	43	51
24	27	37	
21	25	34	
36	38		
24			

Null hypothesis: $\mu_1 = \mu_2 = \mu_3 = \mu_4$

Anova: Single Factor

Research hypothesis: $\mu_1 \neq \mu_2 \neq \mu_3 \neq \mu_4$

SUMMARY

Groups	*Count*	*Sum*	*Average*	*Variance*
18 - 30	9	249	27.67	74.00
31 - 50	8	248	31.00	79.43
51 - 65	7	289	41.29	131.90
66 +	5	279	55.80	87.70

ANOVA

Source of Variation	*SS*	*df*	*MS*	*F*	*P-value*	*F crit*
Between Groups	2965.56	3	988.52	10.79	9.81E-05	2.99
Within Groups	2290.23	25	91.61			
Total	5255.79	28				

Age 18-30 vs. Age 51-65

1/age 18-30 + 1/age 51-65	0.25
s.e.	4.82
ANOVA t	-2.82

Fig. A.23 Answer to Chap. 8: Practice Problem #2

Chapter 8: Practice Problem #2 (continued)

1. Null hypothesis : $\mu_1 = \mu_2 = \mu_3 = \mu_4$
   Research hypothesis : $\mu_1 \neq \mu_2 \neq \mu_3 \neq \mu_4$
2. $MS_b = 988.52$
3. $MS_w = 91.61$
4. $F = 10.79$
5. critical $F = 2.99$
6. Since the F-value of 10.79 is greater than the critical F value of 2.99, we reject the null hypothesis and accept the research hypothesis.
7. There was a significant difference in voting patterns between the different age groups.
8. Null hypothesis : $\mu_1 = \mu_3$
   Research hypothesis : $\mu_1 \neq \mu_3$
9. 27.67 %
10. 41.29 %
11. degrees of freedom $= 29 - 4 = 25$
12. critical $t = 2.060$
13. $\text{s.e.}_{\text{ANOVA}} = \text{SQRT}(MS_w \times \{1/9 + 1/7\}) = \text{SQRT}(91.61 \times \{0.111 + 0.143\})$
    $= \text{SQRT}(23.27) = 4.82$
14. ANOVA t $= (27.67 - 41.29) / 4.82 = -2.83$
15. Since the absolute value of -2.83 is greater than the critical t of 2.060, we reject the null hypothesis and accept the research hypothesis.
16. A significantly higher percentage of voters age 51–65 voted in Illinois in the last Presidential election than age 18–30 (41 % vs. 28 %).

Chapter 8: Practice Problem #3 Answer (see Fig. A.24)

Chapter 8: Practice Problem #3: Answer

U.S. DEFENSE SPENDING AS A PERCENT OF THE NATIONAL BUDGET

Question: Does political leaning affect attitude toward U.S. defense spending?

Item #1: Do you consider yourself to be liberal, moderate, or conservative in your political views?

Item #3: What is your attitude toward U.S. government spending on defense as a percentage of the U.S. national budget?

1	2	3	4	5	6	7	8	9	10
It should be decreased				It should remain about the same					It should be increased

Political preference

Liberal	Moderate	Conservative
3	4	4
5	5	7
4	6	6
6	4	8
7	5	5
5	3	9
4	7	10
3	6	7
2	4	5
1		6
		8

Null hypothesis: $\mu_1 = \mu_2 = \mu_3$

Research hypothesis: $\mu_1 \neq \mu_2 \neq \mu_3$

Anova: Single Factor

SUMMARY

Groups	Count	Sum	Average	Variance
Liberal	10	40	4.00	3.33
Moderate	9	44	4.89	1.61
Conservative	11	75	6.82	3.36

ANOVA

Source of Variation	SS	df	MS	F	P-value	F crit
Between Groups	43.77	2	21.89	7.72	0.002	3.35
Within Groups	76.53	27	2.83			
Total	120.30	29				

Moderates vs. Conservatives

1/n Moderates + 1/n Conservatives	0.20
s.e.	0.76
ANOVA t	-2.55

Fig. A.24 Answer to Chap. 8: Practice Problem #3

Chapter 8: Practice Problem #3 (continued)

Let Group 1 = Liberal, Group 2 = Moderate, and Group 3 = Conservative

1. Null hypothesis : $\mu_1 = \mu_2 = \mu_3$
   Research hypothesis : $\mu_1 \neq \mu_2 \neq \mu_3$
2. $MS_b = 21.89$
3. $MS_w = 2.83$
4. F = 7.73
5. critical F = 3.35
6. Since the F-value of 7.73 is greater than the critical F value of 3.35, we reject the null hypothesis and accept the research hypothesis.
7. There was a significant difference in attitude toward U.S. Defense spending between the three political groups.
8. Null hypothesis : $\mu_2 = \mu_3$
   Research hypothesis : $\mu_2 \neq \mu_3$
9. 4.89
10. 6.82
11. degrees of freedom = 30 − 3 = 27
12. critical t = 2.052
13. $s.e._{ANOVA} = SQRT(MS_w \times \{1/9+1/11\}) = SQRT(2.83 \times \{0.111 + 0.091\}) = SQRT(0.57) = 0.76$
14. ANOVA t = (4.89 − 6.82) /0.76 = −2.54
15. Since the absolute value of −2.54 is greater than the critical t of 2.052, we reject the null hypothesis and accept the research hypothesis.
16. Conservatives were significantly more likely to feel that U.S. government spending as a percentage of the U.S. national budget should be increased than moderates (6.82 vs. 4.89).

Appendix B: Practice Test

Chapter 1: Practice Test

Suppose that you have been asked by the manager of the Webster Groves Subaru dealer in St. Louis to analyze the data from a recent survey of its customers. Subaru of America mails a "SERVICE EXPERIENCE SURVEY" to customers who have recently used the Service Department for their car. Let's try your Excel skills on Item #10e of this survey (see Fig. B.1).

Item #10e:	"Your overall rating of the quality of work performed on your vehicle."								
1	2	3	4	5	6	7	8	9	10
Unacceptable									Extraordinary
	RATING								
	8								
	5								
	6								
	5								
	4								
	8								
	7								
	7								
	8								
	6								
	7								
	5								
	4								
	8								
	7								
	5								
	7								
	5								
	7								
	6								

Fig. B.1 Worksheet Data for Chap. 1 Practice Test (Practical Example)

(a) Create an Excel table for these data, and then use Excel to the right of the table to find the sample size, mean, standard deviation, and standard error of the mean for these data. Label your answers, and round off the mean, standard deviation, and standard error of the mean to two decimal places.

(b) Save the file as: SUBARU8

Chapter 2: Practice Test

Suppose that you wanted to do a personal interview with a random sample of 12 of a school district's 42 high school science teachers as part of a curriculum revision project.

(a) Set up a spreadsheet of frame numbers for these science teachers with the heading: FRAME NUMBERS
(b) Then, create a separate column to the right of these frame numbers which duplicates these frame numbers with the title: Duplicate frame numbers.
(c) Then, create a separate column to the right of these duplicate frame numbers called RAND NO. and use the *=RAND()* function to assign random numbers to all of the frame numbers in the duplicate frame numbers column, and change this column format so that three decimal places appear for each random number.
(d) Sort the *duplicate frame numbers and random numbers* into a random order.
(e) Print the result so that the spreadsheet fits onto one page.
(f) Circle on your printout the I.D. number of the first 12 science teachers that you would interview in science curriculum revision project.
(g) Save the file as: RAND15

Important note: Note that everyone who does this problem will generate a different random order of teacher ID numbers since Excel assign a different random number each time the RAND() command is used. For this reason, the answer to this problem given in this Excel Guide will have a completely different sequence of random numbers from the random sequence that you generate. This is normal and what is to be expected.

Chapter 3: Practice Test

Webster University, with headquarters in St. Louis, Missouri USA, has over 60 sites where students can take courses, including sites in four European campuses and four sites in Asia for its 17,000 students. Each term, students complete a Course Feedback form at the end of the course and confidential results are given to the instructors several weeks after the course is completed. Suppose that you have been asked to analyze the data for classes in St. Louis for the previous term, and to test your Excel skills, you have selected a random sample of students from one of the courses. The hypothetical data for Item #7 appear in Fig. B.2.

WEBSTER UNIVERSITY				
School of Business and Technology				
Item #7:	"Instructor's ability to explain concepts clearly."			
	1	2	3	4
	Very Effective	Effective	Ineffective	Very Ineffective
	Results:			
	3			
	1			
	3			
	1			
	2			
	1			
	3			
	2			
	1			
	4			
	1			
	2			
	1			
	1			
	2			
	3			
	1			
	2			

Fig. B.2 Worksheet Data for Chap. 3 Practice Test (Practical Example)

(a) Create an Excel table for these data, and use Excel to the right of the table to find the sample size, mean, standard deviation, and standard error of the mean for these data. Label your answers, and round off the mean, standard deviation, and standard error of the mean to two decimal places in number format.
(b) By hand, write the null hypothesis and the research hypothesis on your printout.
(c) Use Excel's *TINV function* to find the 95 % confidence interval about the mean for these data. Label your answers. Use two decimal places for the confidence interval figures in number format.
(d) On your printout, draw a diagram of this 95 % confidence interval by hand, including the reference value.
(e) On your spreadsheet, enter the *result.*
(f) On your spreadsheet, enter the *conclusion in plain English.*
(g) Print the data and the results so that your spreadsheet fits onto one page.
(h) Save the file as: Webster5

Chapter 4: Practice Test

Suppose that you have been asked by one of your professors to act as a research assistant on a survey project of undergraduates at a major university. The professor is interested in studying the attitudes of students on a variety of issues, one of them being capital punishment of convicted criminals. The professor has developed a survey that can be mailed to current students, and item #14 on the survey asks the students for their opinion on capital punishment in terms of the death penalty. Assume that the data have been collected, and you decide to take a random sample of student responses to this one item so that you can test out your Excel skills on a small sample instead of the much larger sample of students involved in this survey. A random sample of the hypothetical data for this one item is given in Fig. B.3.

DEATH PENALTY ATTITUDE ITEM ON SURVEY							
Item #14	"How do you feel about capital punishment in the death penalty?"						
	1	2	3	4	5	6	7
	oppose						favor
	Item #14						
	2						
	5						
	4						
	7						
	3						
	4						
	2						
	1						
	3						
	2						
	4						
	1						
	5						
	1						
	2						
	3						
	1						
	2						

Fig. B.3 Worksheet Data for Chap. 4 Practice Test (Practical Example)

(a) Write the null hypothesis and the research hypothesis on your spreadsheet.
(b) Create a spreadsheet for these data, and then use Excel to find the sample size, mean, standard deviation, and standard error of the mean to the right of the data set. Use number format (two decimal places) for the mean, standard deviation, and standard error of the mean.
(c) Type the *critical t* from the t-table in Appendix E onto your spreadsheet, and label it.
(d) Use Excel to compute the t-test value for these data (use three decimal places) and label it on your spreadsheet.
(e) Type the *result* on your spreadsheet, and then type the *conclusion in plain English* on your spreadsheet.
(f) Save the file as: Capital3

Chapter 5: Practice Test

Massachusetts Mutual Financial Group (2010) placed a full-page color ad in *The Wall Street Journal* in which it used a male model hugging a 2-year old daughter. The ad had the headline and sub-headline:

WHAT IS THE SIGN OF A GOOD DECISION?

It's knowing your life insurance can help provide income for retirement. And peace of mine until you get there.

Since the majority of the subscribers to *The Wall Street Journal* are men, an interesting research question would be the following:

Research question: "Does the gender of the model affect adult men's willingness to learn more about how life insurance can provide income for retirement?"

Suppose that you have shown two groups of adult males (ages 25–44) a mockup of an ad such one group of males saw the ad with a male model, while another group of males saw the identical ad except that it had a female model in the ad. (You randomly assigned these males to one of the two experimental groups.) The two groups were kept separate during the experiment and could not interact with one another.

At the end of a 1-hour discussion of the mockup ad, the respondents were asked the question given in Fig. B.4.

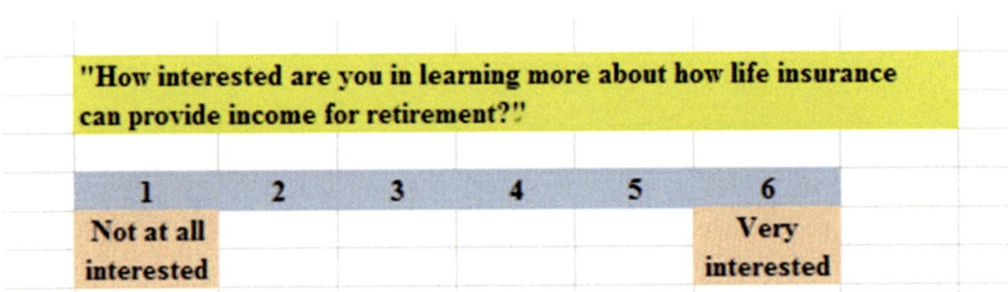

Fig. B.4 Survey Item for a Mockup Ad (Practical Example)

MASS MUTUAL FINANCIAL GROUP					
Item:	**"How interested are you in learning more about how life insurance can provide income for retirement?"**				
1	**2**	**3**	**4**	**5**	**6**
Not at all interested					**Very interested**

Male model	Female model
3	4
2	6
4	5
5	3
1	4
6	6
2	6
4	5
3	3
5	5
2	4
4	3
3	5
5	4
1	6
2	5
3	5
1	6
4	4
5	6
6	3
2	4
3	6
1	5
4	6
3	4
5	4

Fig. B.5 Worksheet Data for Chap. 5 Practice Test (Practical Example)

The resulting data for this one item appear in Fig. B.5.

(a) Write the null hypothesis and the research hypothesis.
(b) Create an Excel table that summarizes these data.
(c) Use Excel to find the standard error of the difference of the means.
(d) Use Excel to perform a *two-group t-test*. What is the value of *t* that you obtain (use two decimal places)?
(e) On your spreadsheet, type the *critical value of t* using the t-table in Appendix E.
(f) Type the *result* of the test on your spreadsheet.
(g) Type your *conclusion in plain English* on your spreadsheet.
(h) Save the file as: lifeinsur3
(i) Print the final spreadsheet so that it fits onto one page.

Chapter 6: Practice Test

Is there a relationship between seniority (i.e., the number of years served in the U.S. Congress) and the margin of victory (i.e. the "closeness" of the race in terms of the number of percentage points the winner had over the second place finisher) in that Congress person's last election to Congress? You have decided to use seniority as the independent variable (predictor) and margin of victory as the dependent variable (criterion). Does seniority increase as the margin of victory increases? Shively (2009) used hypothetical data in an example in his book to show how you could test this relationship. Use simple linear regression for the hypothetical data given in Fig. B.6:

RELATIONSHIP BETWEEN SENIORITY AND MARGIN OF VICTORY FOR U.S. CONGRESSIONAL DISTRICTS

X = No. of years served in the U.S. Congress (seniority)

Y = Margin of victory percent

SENIORITY	MARGIN OF VICTORY
2	1
4	3
8	5
12	6
15	3
29	8
18	4
14	8
12	6
16	12
32	2
26	16
20	18

Fig. B.6 Worksheet Data for Chap. 6 Practice Test (Practical Example)

Create an Excel spreadsheet, and enter the data.

(a) create an *XY scatterplot* of these two sets of data such that:

- top title: RELATIONSHIP BETWEEN SENIORITY AND MARGIN OF VICTORY FOR U.S. CONGRESSIONAL DISTRICTS
- x-axis title: SENIORITY
- y-axis title: MARGIN OF VICTORY
- move the chart below the table
- re-size the chart so that it is 7 columns wide and 25 rows long
- delete the legend
- delete the gridlines

(b) Create the *least-squares regression line* for these data on the scatterplot.
(c) Use Excel to run the regression statistics to find the *equation for the least-squares regression line* for these data and display the results below the chart on your spreadsheet. Use Excel to add the regression line to the chart. Use number format (two decimal places) for the correlation and for the coefficients

Print *just the input data and the chart* so that this information fits onto one page in portrait format.

Then, print *just the regression output table* on a separate page so that it fits onto that separate page in portrait format.

By hand:

(d) Circle and label the value of the *y-intercept* and the *slope* of the regression line on your printout.
(e) Write the regression equation *by hand* on your printout for these data (use two decimal places for the y-intercept and the slope).
(f) Circle and label the *correlation* between the two sets of scores in the regression analysis summary output table on your printout.
(g) Underneath the regression equation you wrote by hand on your printout, use the regression equation to predict the margin of victory you would expect for a seniority of 12 years.
(h) save the file as: Congress3

Chapter 7: Practice Test

Suppose that you have been hired by a political pollster to see how well you could predict the percentage of Democratic legislators in the state legislatures of the U.S. You decide to use the percent of registered voters in a state who say they are Democrats as one predictor, and the percent of the state's registered voters who say they are "liberal" in their political views as a second predictor, and the percent of state legislators who are Democrats as the criterion in a multiple regression situation. To check your skill in Excel, you have selected a small random sample of states and recorded the information on each state given in the hypothetical table in Fig. B.7 before you do this analysis using all 50 states of the U.S.

POLITICAL AFFILIATION AND ITS EFFECT ON STATE LEGISLATORS

Question: "How well does the percent of registered voters in a state who say they are liberal in their political views, and the percent of registered voters in that state who say they are Democrats, predict the percent of state legislators who are Democrats?

Y = % of state legislators who are Democrats

X1 = % of state registered voters who say they are Democrats

X2 = % of state registered voters who say they are "liberal" in their political views

Y	X1	X2
Democratic legislators	Democrats registered	Liberal
46	43	50
44	44	46
51	45	45
48	42	43
49	38	41
52	39	42
47	41	38
45	42	46
54	39	48
52	43	50
48	41	47
49	48	48
47	49	46

Fig. B.7 Worksheet Data for Chap. 7 Practice Test (Practical Example)

(a) create an Excel spreadsheet using Democratic legislators as the criterion (Y), and the other variables as the two predictors of this criterion (X_1 = Democrats registered, X_2 = Liberal).

(b) Use Excel's *multiple regression* function to find the relationship between these three variables and place the SUMMARY OUTPUT below the table.

(c) Use number format (two decimal places) for the multiple correlation on the Summary Output, and use two decimal places for the coefficients in the SUMMARY OUTPUT.
(d) Save the file as: Political3
(e) Print the table and regression results below the table so that they fit onto one page.

Answer the following questions using your Excel printout:

1. What is the multiple correlation R_{xy}?
2. What is the y-intercept a?
3. What is the coefficient for percent of registered voters who are Democrats: b_1?
4. What is the coefficient for percent of registered voters who say they are “liberal”: b_2?
5. What is the multiple regression equation?
6. Predict the percent of Democratic legislators you would expect for a state where 41 % of the registered voters said they were Democrats and 48 % of its registered voters said they were “liberal.”

(f) Now, go back to your Excel file and create a correlation matrix for these three variables, and place it underneath the SUMMARY OUTPUT.
(g) Save this file as: Political3
(h) Now, print out *just this correlation matrix* on a separate sheet of paper.

Answer to the following questions using your Excel printout. (Be sure to include the plus or minus sign for each correlation):

7. What is the correlation between % registered Democrats vs. % Democratic legislators?
8. What is the correlation between % liberal registered voters vs. % Democratic legislators?
9. What is the correlation between % registered Democrats vs. % liberal registered voters?
10. Discuss which of the two predictors is the better predictor of % Democratic legislators.
11. Explain in words how much better the two predictor variables combined predict % Democratic legislators than the better single predictor by itself.

Chapter 8: Practice Test

Suppose that you have been asked to analyze data from a study in which the data for the percent of registered voters in Missouri who voted in the last Congressional election were broken down by Congressional district, and then again by the educational attainment level of these voters in each district. What is the relationship between voting patterns and the educational attainment of the voters in Missouri? Your task is to determine if there was a relationship during the last Congressional election using the hypothetical data in Fig. B.8. You have selected a small sample of Congressional districts, and because the collection of these data took longer than you anticipated, you have a different number of data points for the four educational attainment groups, but you want to test your Excel skills on the data that you have collected so far in any case before you complete the larger data collection part of this study.

VOTING PATTERNS BASED ON EDUCATIONAL ACHIEVEMENT

Question: "How is educational achievement related to voting patterns?

Percent of registered voters in Missouri who voted in the last Congressional election

< high school diploma	h.s. diploma	college degree	advanced degree
18	21	33	45
22	24	35	57
21	26	34	54
16	18	32	56
17	19	36	71
19	23	37	64
24	26	51	54
17	29		67
16			75
			67

Fig. B.8 Worksheet Data for Chap. 8 Practice Test (Practical Example)

(a) Enter these data on an Excel spreadsheet.
(b) On your spreadsheet, write the null hypothesis and the research hypothesis for these data
(c) Perform a *one-way ANOVA test* on these data, and show the resulting ANOVA table underneath the input data for the four types of educational attainment.
(d) If the F-value in the ANOVA table is significant, create an Excel formula to compute the ANOVA t-test comparing education less than high school versus College degree, and show the results below the ANOVA table on the spreadsheet (put the standard error and the ANOVA t-test value on separate lines of your spreadsheet, and use two decimal places for each value)
(e) Print out the resulting spreadsheet so that all of the information fits onto one page
(f) On your printout, label by hand the MS (between groups) and the MS (within groups)
(g) Circle and label the value for F on your printout for the ANOVA of the input data
(h) Label by hand on the printout the mean for less than h.s. and the mean for College degree that were produced by your ANOVA formulas
(i) Save the spreadsheet as: Voting3

On a separate sheet of paper, now do the following by hand:

(j) find the critical value of F in the ANOVA Single Factor results table
(k) write a summary of the *result* of the ANOVA test for the input data
(l) write a summary of the *conclusion* of the ANOVA test in plain English for the input data
(m) write the null hypothesis and the research hypothesis comparing less than h.s. versus College degree
(n) compute the degrees of freedom for the *ANOVA t-test* by hand for these two groups.
(o) write the *critical value of t* for the ANOVA t-test using the table in Appendix E.
(p) write a summary of the *result* of the ANOVA t-test
(q) write a summary of the *conclusion* of the ANOVA t-test in plain English

References

Mass Mutual Financial Group. What is the Sign of a Good Decision? (Advertisement) *The Wall Street Journal*, September 29, 2010, p. A22.

Shively, W.P. The Craft of Political Research (7th ed.). Upper Saddle River, NJ: Prentice Hall/ Pearson, 2009.

Appendix C: Answers to Practice Test

Practice Test Answer: Chap. 1 (see Fig. C.1)

Practice Test Answer: Chapter 1 (see Fig.C.1)

Item #10e: **"Your overall rating of the quality of work performed on your vehicle."**

1	2	3	4	5	6	7	8	9	10
Unacceptable									Extraordinary

RATING		
8		
5		
6	n	20
5		
4		
8	Mean	6.25
7		
7		
8	STDEV	1.33
6		
7		
5	s.e.	0.30
4		
8		
7		
5		
7		
5		
7		
6		

Fig. C.1 Practice Test Answer to Chap. 1 Problem

Practice Test Answer: Chap. 2 (see Fig. C.2)

Practice Test Answer: Chapter 2			
	FRAME NUMBERS	Duplicate frame numbers	RAND NO.
	1	8	0.957
	2	22	0.907
	3	31	0.048
	4	42	0.552
	5	4	0.959
	6	29	0.610
	7	3	0.876
	8	21	0.363
	9	37	0.078
	10	17	0.799
	11	34	0.890
	12	25	0.358
	13	10	0.663
	14	41	0.179
	15	30	0.690
	16	36	0.954
	17	13	0.582
	18	15	0.063
	19	20	0.178
	20	14	0.155
	21	9	0.836
	22	12	0.016
	23	38	0.531
	24	26	0.295
	25	1	0.450
	26	5	0.267
	27	35	0.651
	28	28	0.154
	29	24	0.832
	30	32	0.988
	31	27	0.985
	32	19	0.562
	33	6	0.554
	34	39	0.382
	35	2	0.237
	36	18	0.003
	37	7	0.055
	38	11	0.623
	39	16	0.662
	40	40	0.457
	41	33	0.916
	42	23	0.666

Fig. C.2 Practice Test Answer to Chap. 2 Problem

Practice Test Answer: Chap. 3 (see Fig. C.3)

Practice Test Answer: Chapter 3

WEBSTER UNIVERSITY

School of Business and Technology

Item #7: **"Instructor's ability to explain concepts clearly."**

1	2	3	4
Very Effective	Effective	Ineffective	Very Ineffective

Results:

Results		
3		
1	Null hypothesis:	μ = 2.5
3		
1		
2	Research hypothesis:	μ ≠ 2.5
1		
3		
2	n	18
1		
4	mean	1.89
1		
2	stdev	0.96
1		
1	s.e.	0.23
2		
3		
1	95% confidence interval	
2		

lower limit	1.41
upper limit	2.37

------ 1.41---------- 1.89 -------------- 2.37 -------------- 2.5

lower limit	mean	upper limit	Ref. Value

Result: Since the reference value is outside the confidence interval, we reject the null hypothesis and accept the research hypothesis

Conclusion: Students in this course rated the instructor's ability to explain concepts clearly as significantly effective.

Fig. C.3 Practice Test Answer to Chap. 3 Problem

Practice Test Answer: Chap. 4 (see Fig. C.4)

Practice Test Answer: Chapter 4							
DEATH PENALTY ATTITUDE ITEM ON SURVEY							
Item #14	"How do you feel about capital punishment in the death penalty?"						
	1	2	3	4	5	6	7
	oppose						favor
	Item #14		n	18			
	2					H_0	$\mu = 4$
	5		mean	2.89			
	4					H_1	$\mu \neq 4$
	7						
	3		STDEV	1.68			
	4						
	2						
	1		s.e.	0.40			
	3						
	2						
	4		t-test	-2.81			
	1						
	5						
	1		critical t	2.11			
	2						
	3						
	1		Result:	Since the absolute value of − 2.81 is greater than the critical t of 2.11, we reject the null hypothesis and accept the research hypothesis			
	2						
			Conclusion:	Undergraduates at State U. significantly oppose the death penalty in capital punishment			

Fig. C.4 Practice Test Answer to Chap. 4 Problem

Practice Test Answer: Chap. 5 (see Fig. C.5)

Practice Test Answer: Chapter 5

MASS MUTUAL FINANCIAL GROUP

Item: **"How interested are you in learning more about how life insurance can provide income for retirement?"**

1	2	3	4	5	6
Not at all interested		3.30		4.70	Very interested

Male model	Female model
3	4
2	6
4	5
5	3
1	4
6	6
2	6
4	5
3	3
5	5
2	4
4	3
3	5
5	4
1	6
2	5
3	5
1	6
4	4
5	6
6	3
2	4
3	6
1	5
4	6
3	4
5	4

Group	n	Mean	STDEV
1 Male model	27	3.30	1.54
2 Female model	27	4.70	1.07

Null hypothesis: $\mu_1 = \mu_2$

Research hypothesis: $\mu_1 \neq \mu_2$

1/n1 + 1/n2	0.07
(n1 - 1) x S1 squared	61.63
(n2 - 1) x S2 squared	29.63
n1 + n2 - 2 (degrees of freedom)	52
s.e.	0.36
critical t	1.96
t-test	-3.90

Result: Since the absolute value of - 3.90 is greater than the critical t of 1.96, we reject the null hypothesis and accept the research hypothesis.

Conclusion: Adult men (ages 25-44) were significantly more interested in learning more about how life insurance can provide income for retirement when a female model was used than when a male model was used in the ad (4.70 vs. 3.30)

Fig. C.5 Practice Test Answer to Chap. 5 Problem

Practice Test Answer: Chap. 6 (see Fig. C.6)

Practice Test Answer: Chapter 6

RELATIONSHIP BETWEEN SENIORITY AND MARGIN OF VICTORY FOR U.S. CONGRESSIONAL DISTRICTS

X = No. of years served in the U.S. Congress (seniority)

Y = Margin of victory percent

SENIORITY	MARGIN OF VICTORY
2	1
4	3
8	5
12	6
15	3
29	8
18	4
14	8
12	6
16	12
32	2
26	16
20	18

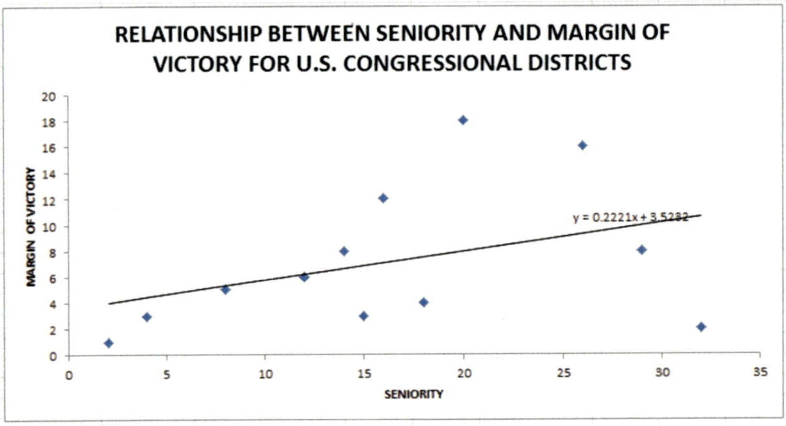

SUMMARY OUTPUT

Regression Statistics	
Multiple R	0.38
R Square	0.14
Adjusted R Square	0.07
Standard Error	5.12
Observations	13

ANOVA

	df	*SS*	*MS*	*F*	*Significance F*
Regression	1	48.64	48.64	1.86	0.20
Residual	11	288.28	26.21		
Total	12	336.92			

	Coefficients	*Standard Error*	*t Stat*	*P-value*	*Lower 95%*
Intercept	3.52	2.97	1.19	0.26	-3.01
X Variable 1	0.22	0.16	1.36	0.20	-0.14

Fig. C.6 Practice Test Answer to Chap. 6 Problem

Practice Test Answer: Chap. 6: (continued)

(d) a = y-intercept = 3.52
  b = slope = +0.22

(e) $Y = a + bX$
  $Y = 3.52 + 0.22X$

(f) r = correlation = +.38

(g) $Y = 3.52 + 0.22(12)$
  $Y = 3.52 + 2.64$
  $Y = 6.16 = 6\%$ margin of victory

Practice Test Answer: Chap. 7 (see Fig. C.7)

Practice Test Answer: Chapter 7

POLITICAL AFFILIATION AND ITS EFFECT ON STATE LEGISLATORS

Question: "How well does the percent of registered voters in a state who say they are liberal in their political views, and the percent of registered voters in that state who say they are Democrats, predict the percent of state legislators who are Democrats?

Y = % of state legislators who are Democrats

X1 = % of state registered voters who say they are Democrats

X2 = % of state registered voters who say they are "liberal" in their political views

Y	X1	X2
Democratic legislators	Democrats registered	Liberal
46	43	50
44	44	46
51	45	45
48	42	43
49	38	41
52	39	42
47	41	38
45	42	46
54	39	48
52	43	50
48	41	47
49	48	48
47	49	46

SUMMARY OUTPUT

Regression Statistics	
Multiple R	0.38
R Square	0.14
Adjusted R Square	-0.03
Standard Error	3.00
Observations	13

ANOVA

	df	*SS*	*MS*	*F*	*Significance F*
Regression	2	15.11	7.56	0.84	0.46
Residual	10	89.96	9.00		
Total	12	105.08			

	Coefficients	*Standard Error*	*t Stat*	*P-value*	*Lower 95%*
Intercept	54.71	13.36	4.09	0.00	24.94
Democrats registered	-0.36	0.28	-1.26	0.24	-0.99
Liberal	0.20	0.27	0.77	0.46	-0.39

	Democratic legislators	*Democrats registered*	*Liberal*
Democratic legislators	1		
Democrats registered	-0.31	1	
Liberal	0.09	0.39	1

Fig. C.7 Practice Test Answer to Chap. 7 Problem

Practice Test Answer: Chap. 7 (continued)

1. $R_{xy} = +0.38$
2. y-intercept a $= 54.71$
3. $b_1 = -0.36$
4. $b_2 = 0.20$
5. $Y = a + b_1X_1 + b_2X_2$
   $Y = 54.71 - 0.36X_1 + 0.20X_2$
6. $Y = 54.71 - 0.36(41) + 0.20(48)$
   $Y = 54.71 - 14.76 + 9.6$
   $Y = 49.55$
   $Y = 50\%$
7. $-.31$
8. $+.09$
9. $+.39$
10. Percent Democrats registered is a better predictor of percent Democratic legislators ($r = -.31$) than is a liberal political view ($r = +.09$).
11. The two predictors combined predict the percent of Democratic legislators much better than either single predictor $\left(R_{xy} = +.38\right)$.

Practice Test Answer: Chap. 8 (see Fig. C.8)

Practice Test Answer: Chapter 8

VOTING PATTERNS BASED ON EDUCATIONAL ACHIEVEMENT

Question: "How is educational achievement related to voting patterns?

Percent of registered voters in Missouri who voted in the last Congressional election

<high school diploma	h.s. diploma	college degree	advanced degree
18	21	33	45
22	24	35	57
21	26	34	54
16	18	32	56
17	19	36	71
19	23	37	64
24	26	51	54
17	29		67
16			75
			67

Anova: Single Factor

SUMMARY

Groups	Count	Sum	Average	Variance
<high school diploma	9	170	18.89	8.11
h.s. diploma	8	186	23.25	14.21
college degree	7	258	36.86	41.81
advanced degree	10	610	61.00	85.78

ANOVA

Source of Variation	SS	df	MS	F	P-value	F crit
Between Groups	10190.75	3	3396.92	85.84	8.05E-15	2.92
Within Groups	1187.25	30	39.57			
Total	11378.00	33				

<high school diploma vs. College degree

1/ n <h.s. + 1/n college	0.25
s.e.	3.17
ANOVA t	-5.67

Fig. C.8 Practice Test Answer to Chap. 8 Problem

Let Group 1 = less than h.s., Group 2 = h.s., Group 3 = college, and Group 4 = advanced.

(b) $H_0 : \mu_1 = \mu_2 = \mu_3 = \mu_4$
$H_1 : \mu_1 \neq \mu_2 \neq \mu_3 \neq \mu_4$

(f) $MS_b = 3396.92$ and $MS_w = 39.57$

(g) $F = 85.84$

(h) Mean of less than h.s. = 18.89 and Mean of College = 36.86

(j) critical F = 2.92

(k) Result: Since 85.84 is greater than 2.92, we reject the null hypothesis and accept the research hypothesis

(l) Conclusion: There was a significant difference in the registered voters in Missouri who voted in the last Congressional election based on the educational achievement level.

(m) $H_0 : \mu_1 = \mu_3$
$H_1 : \mu_1 \neq \mu_3$

(n) $df = n_{TOTAL} - k = 34 - 4 = 30$

(o) critical t = 2.042

(p) Result: Since the absolute value of −5.67 is greater than the critical t of 2.042, we reject the null hypothesis and accept the research hypothesis.

(q) A significantly higher percentage of registered voters in Missouri voted in the last Congressional election with College degrees than with less than a high school education (37 % vs. 19 %).

Appendix D: Statistical Formulas

Mean

$$\overline{X} = \frac{\Sigma X}{n}$$

Standard Deviation

$$\text{STDEV} = S = \sqrt{\frac{\Sigma(X - \overline{X})^2}{n - 1}}$$

Standard error of the mean

$$\text{s.e.} = S_{\overline{X}} = \frac{s}{\sqrt{n}}$$

Confidence interval about the mean

$$\overline{X} \pm t S_{\overline{X}}$$

$$\text{where } S_{\overline{X}} = \frac{S}{\sqrt{n}}$$

One-group t-test

$$t = \frac{\overline{X} - \mu}{s_{\overline{X}}}$$

$$\text{where } S_{\overline{X}} = \frac{s}{\sqrt{n}}$$

Two-group t-test

(a) when both groups have a sample size greater than 30

$$t = \frac{\overline{X}_1 - \overline{X}_2}{S_{\overline{X}_1 - \overline{X}_2}}$$

$$\text{where } S_{\overline{X}_1 - \overline{X}_2} = \sqrt{\frac{S_1{}^2}{n_1} + \frac{S_2{}^2}{n_2}}$$

$$\text{and where } df = n_1 + n_2 - 2$$

(b) when one or both groups have a sample size less than 30

$$t = \frac{\overline{X}_1 - \overline{X}_2}{S_{\overline{X}_1 - \overline{X}_2}}$$

$$\text{where } S_{\overline{X}_1 - \overline{X}_2} = \sqrt{\frac{(n_1 - 1)S_1{}^2 + (n_2 - 1)S_2{}^2}{n_1 + n_2 - 2}\left(\frac{1}{n_1} + \frac{1}{n_2}\right)}$$

$$\text{and where } df = n_1 + n_2 - 2$$

Correlation

$$r = \frac{\frac{1}{n-1}\Sigma\left(X - \overline{X}\right)\left(Y - \overline{Y}\right)}{S_x S_y}$$

where S_x = standard deviation of X
and where S_y = standard deviation of Y

Simple linear regression

$Y = a + bX$
where a = y-intercept and b = slope of the line

Multiple regression equation

$Y = a + b_1 X_1 + b_2 X_2 + b_3 X_3 + \text{etc.}$
where a = y-intercept

One-way ANOVA F-test

$F = MS_b / MS_w$

ANOVA t-test

$$ANOVA\, t = \frac{\overline{X}_1 - \overline{X}_2}{s.e._{ANOVA}}$$

where $s.e._{ANOVA} = \sqrt{MS_w\left(\frac{1}{n_1} + \frac{1}{n_2}\right)}$
and where $df = n_{TOTAL} - k$
where $n_{TOTAL} = n_1 + n_2 + n_3 + \text{etc.}$
and where k = the number of groups

Appendix E: t-Table

Critical t-values needed for rejection of the null hypothesis (see Fig. E.1)

Fig. E.1 Critical t-values Needed for Rejection of the Null Hypothesis

sample size n	degrees of freedom df	critical t
10	9	2.262
11	10	2.228
12	11	2.201
13	12	2.179
14	13	2.160
15	14	2.145
16	15	2.131
17	16	2.120
18	17	2.110
19	18	2.101
20	19	2.093
21	20	2.086
22	21	2.080
23	22	2.074
24	23	2.069
25	24	2.064
26	25	2.060
27	26	2.056
28	27	2.052
29	28	2.048
30	29	2.045
31	30	2.042
32	31	2.040
33	32	2.037
34	33	2.035
35	34	2.032
36	35	2.030
37	36	2.028
38	37	2.026
39	38	2.024
40	39	2.023
infinity	infinity	1.960

Index

© Springer International Publishing Switzerland 2016
T.J. Quirk, *Excel 2016 for Social Science Statistics*, Excel for Statistics,
DOI 10.1007/978-3-319-39711-5

Made in United States
Orlando, FL
04 May 2026

81141782R00167